Revise A2 Biology for OCR

Jennifer Gregory, Ianto Stevens and Richard Fosbery

Heinemann Educational Publishers
Halley Court, Jordan Hill, Oxford, OX2 8EJ
Part of Harcourt Education

Heinemann is a registered trademark of Harcourt Education Limited

First published 2001

13-digit ISBN: 978 0 435583 29 3

10 09 08 07
10 9 8 7 6 5 4

Commissioning/development editor Paddy Gannon

Edited by Eva Fairnell

Designed and typeset by Saxon Graphics Ltd, Derby

Printed and bound in the UK by Ashford Colour Press

Acknowledgements
The authors and publishers would like to thank Macmillan Magazines Limited and
Paul D Boyer for permission to reproduce from Nature magazine the diagram on
page 5 showing rotation of ATP synthase.

The publishers have made every effort to trace the copyright holders, but if they
have inadvertently overlooked any, they will be pleased to make the necessary
arrangements at the first opportunity.

Contents

Introduction: how to use this revision guide

This revision guide is for Module 2804 (Central Concepts) of the OCR A2 Biology course. It also helps with Module 2806/01 (Unifying Concepts in Biology).

The guide follows the sequence of learning outcomes in the specification for Module 2804 (Central Concepts) and has the six sections of the specification:

- Energy and Respiration;
- Photosynthesis;
- Populations and Interactions;
- Meiosis, Genetics and Gene Control;
- Classification, Selection and Evolution;
- Control, Coordination and Homeostasis.

The guide is divided into a number of double-page 'spreads', each of which ends with some Quick check questions. These are designed to test your recall and understanding of the topics you have just read about and to help you with the questions you will find in Module 2804. Try to answer the questions without looking back at the topics. Then check your answers with those given on pages 84 to 86 or with the text.

At the end of each of the six sections, there are questions which resemble closely the types of questions you will find in the examination papers. Many of these questions are synoptic to help you with Module 2806/01 (Unifying Concepts in Biology).

The AS Biology Foundation and Transport modules are recommended prior knowledge for the Central Concepts module. Some of the 'tips' will remind you to revise material from your AS course and alert you to synoptic material that could be examined in 2806/01.

When you are revising, have available a copy of the specification. It is written as learning outcomes such as:

Candidates should be able to:

(a) describe and explain sigmoidal population growth in a bacterial culture.

This tells you exactly what you should learn. As you revise, tick off each learning outcome when you are confident that you understand what it means and can write about it and explain it.

Have fun and good luck with your revision!

OCR A2 Biology: assessment

Here is a plan of the modules you will study for your A2 course.

Module name	Contents of module	Length of exam	Total number of marks in the exam (% of Advanced GCE*)	Distribution of marks in the exam
2804 Central Concepts	Energy and Respiration Photosynthesis Populations and Interactions Meiosis, Genetics and Gene Control Classification, Selection and Evolution Control, Coordination and Homeostasis	1 hour 30 minutes	90 (15%)	structured questions 65 marks extended questions 25 marks
2805/01–05 *One* **of the Options in Biology**	Growth, Development and Reproduction *or* Applications of Genetics *or* Environmental Biology *or* Microbiology and Biotechnology *or* Mammalian Physiology and Behaviour	1 hour 30 minutes	90 (15%)	structured questions 65 marks extended questions 25 marks
2806/01 Unifying Concepts in Biology	You are expected to demonstrate a knowledge and understanding of facts, principles and concepts studied in *both* the AS course and in Central Concepts, and apply them in a context which may be unfamiliar.	1 hour 15 minutes	60 (10%)	structured questions 45 marks extended questions 15 marks
2806/02 *or* 2806/03 Experimental Skills 2	Practical work is not covered in this revision guide. Your school or college will organise this assessment which makes up 10% of the marks for Advanced GCE.			

* Remember, 50% of the marks for Advanced GCE come from the AS course.

N.B. You study *one* of the five Options in Biology (2805)

Question types:

- **Structured questions** require brief answers to several linked parts of a question.
- **Extended questions** require longer answers to a single question.
- Your answers to the extended questions will be used to assess the quality of your **written communication.**
- **Synoptic questions** require you to make and use connections between different areas of the subject, or provide opportunities for you to use skills that permeate the subject, such as data analysis. *No* synoptic questions will be set in Module 2804. The questions in Module 2806/01 will *all* be synoptic. Topics that are considered to be synoptic are listed in the specification on pages 77 and 78.

Module 2804: Central Concepts

Energy and Respiration

The total of all the biochemical reactions needed for an organism to stay alive is its **metabolism**.

metabolism = anabolism + catabolism

Anabolism is the building up of more complex molecules from simpler ones, e.g. the synthesis of nucleic acids and carbohydrates. Enzymes are needed for these syntheses and the complex molecules are needed for growth. Anabolic reactions are energy-consuming.

Catabolism is the enzymic breakdown of complex molecules to simpler ones. It is the opposite of anabolism. The catabolic reactions of respiration yield energy.

✓ *Quick check 1*

All living organisms need a continuous supply of energy to maintain their metabolism. They must absorb either light energy for photosynthesis or chemical potential energy to do the work necessary to stay alive. Such work includes:

- synthesising complex molecules from simpler molecules (anabolic reactions), e.g. polypeptides from amino acids;

- active transport of substances across cell membranes against their concentration gradients, e.g. the activity of the sodium–potassium pump;

- movement of the whole organism by the action of cilia, flagella or muscles, and movement within the organism, e.g. movement of organelles in cells;

- maintenance of body temperature, particularly in mammals and birds where thermal energy is released to maintain the body temperature above that of the environment.

✓ *Quick check 2*

Photosynthesis and respiration

Photosynthesis and respiration are energy transfer processes. Remember the overall equations for them:

photosynthesis $\quad 6CO_2 + 6H_2O \xrightarrow{\text{in the presence of light energy}} C_6H_{12}O_6 + 6O_2$

respiration $\quad C_6H_{12}O_6 + 6O_2 \xrightarrow[\text{energy is released}]{} 6CO_2 + 6H_2O$

Photosynthesis transfers light energy into chemical potential energy, which can then be released for work by the process of **respiration**. Both photosynthesis and respiration involve an important intermediary molecule in this energy transfer: **ATP** (see page 4). In many living organisms, most of the energy transferred to ATP is derived originally from light energy; a few prokaryotes are not dependent on light energy trapped in photosynthesis.

All organisms are made up of macromolecules containing the element carbon: **organic** molecules. All organisms therefore need a supply of carbon to make new organic molecules. Photosynthesis, by **fixing** carbon from carbon dioxide into carbohydrate, is a source of organic molecules.

Autotrophs and heterotrophs

An **autotroph** is an organism that can use an external energy source and simple inorganic molecules to make complex organic molecules. The source of carbon for the organism is carbon dioxide. Autotrophs, e.g. green plants and photosynthetic prokaryotes, that use light energy are called **photoautotrophs**. Some autotrophs use chemical energy sources; these **chemoautotrophs** include the nitrifying bacteria, which obtain their energy by oxidising ammonia to nitrite or nitrite to nitrate.

A **heterotroph** is an organism, e.g. an animal or fungus, that needs to take in complex organic molecules, such as carbohydrates, proteins and lipids, to act as a source of *both* energy and usable carbon compounds.

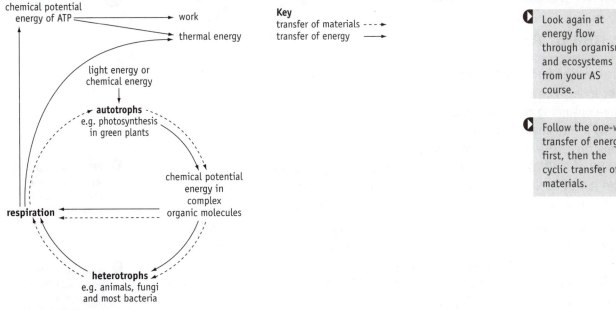

The relationship between photosynthesis and respiration and the transfer of energy and materials from autotroph to heterotroph can be seen in the figure. The energy flows in *one* direction: it cannot be recycled. Materials, e.g. the elements carbon and nitrogen, are recycled.

> **Fixation** of an element is the conversion of an inorganic source of the element to an organic one.

> Review the nitrogen cycle from your AS course.

> ✓ *Quick check 3*

> Look again at energy flow through organisms and ecosystems from your AS course.

> Follow the one-way transfer of energy first, then the cyclic transfer of materials.

> ✓ *Quick check 4*

? Quick check questions

1 Distinguish between anabolism and catabolism.

2 What work must a living organism do to stay alive?

3 Explain the different requirements of an autotroph and a heterotroph.

4 Describe the relationship between photosynthesis and respiration in living organisms.

ATP

Processes in cells that require energy are **linked** to chemical reactions that yield energy by an intermediary molecule: **adenosine triphosphate (ATP)**. Using **one** type of molecule to transfer energy to many **different** energy-requiring processes makes it easier for these processes to be controlled and coordinated. All organisms use ATP as their energy currency: it is a **universal energy currency**.

Structure of ATP

ATP consists of the organic base adenine and the pentose sugar ribose. Together, these make the **nucleoside** adenosine. This is combined with three phosphate groups. ATP is therefore a **nucleotide**.

▶ Look up the structures of the nucleotides in DNA and RNA.

✓ *Quick check 1*

The role of ATP

ATP is the standard intermediary molecule between energy-releasing and energy-consuming reactions. Each cell has only a tiny quantity of ATP in it at any one time. The cell does not import ATP, ADP or AMP. With few exceptions, each cell must produce its own ATP and recycle it very rapidly. Because it is a small water-soluble molecule, it is easily moved from where it is made in a cell to where it is needed.

▶ ATP has so many roles that it is a good topic for the synoptic paper.

ATP can be synthesised from **adenosine diphosphate (ADP)** and an **inorganic phosphate group (P_i)** using energy, and hydrolysed to ADP and phosphate to release energy. This interconversion is all-important in providing energy for a cell. Hydrolysis of the terminal phosphate group of ATP releases 30.5 kJ mol^{-1} of energy for cellular work:

$$\text{ATP} + H_2O \rightleftharpoons \text{ADP} + P_i \pm 30.5 \text{ kJ mol}^{-1}$$

Removing the second phosphate, giving adenosine monophosphate (AMP), also releases 30.5 kJ mol^{-1} of energy, but removing the last phosphate yields only 14.2 kJ mol^{-1}. The energy released does not simply come from these bonds but from the chemical potential energy of the molecule as a whole.

▶ At one time, the bonds attaching the outer two phosphate groups were called 'high energy bonds'. This term should not be used now because it is misleading.

✓ *Quick check 2*

The roles of an ATP molecule include:

- binding to a protein molecule, changing its shape and causing it to fold differently, to produce **movement**, e.g. muscle contraction;
- binding to an enzyme molecule, allowing an energy-requiring reaction to be catalysed;
- transferring a phosphate group to an enzyme, making the enzyme active;
- transferring a phosphate group to an unreactive substrate molecule so that it can react in a specific way, e.g. in glycolysis and the Calvin cycle;
- transferring AMP to an unreactive substrate molecule, producing a reactive intermediate compound, e.g. amino acids before binding to tRNA during protein synthesis;
- binding to a trans-membrane protein so that **active transport** can take place across the membrane.

ATP is never stored. Glucose and fatty acids are short-term energy stores, while glycogen, starch and triglycerides are long-term stores.

The synthesis and subsequent use of ATP involves energy conversion. The energy ultimately ends up as heat. Some organisms are able to control the rate at which their body gains and loses heat.

A **homiotherm** is an organism that can maintain a constant body temperature usually higher than that of the environment.

Synthesis of ATP

✓ *Quick check 3*

There are two ways in which ATP can be synthesised using the energy made available by photosynthesis and respiration.

The energy for the synthesis of *most* ATP comes from electrical potential energy stored as a difference in hydrogen ion (H^+) concentration (a proton gradient) on either side of a phospholipid membrane. This may be in a mitochondrion (see page 9) or in a chloroplast (see page 20).

Note that this process occurs in *both* respiration and photosynthesis.

The inner membrane of the mitochondrion and the thylakoid membrane of the chloroplast are essentially impermeable to hydrogen ions, but the ions flow down their concentration gradient through protein channels spanning the bilayer, as shown in the figure. Part of the protein is the enzyme which synthesises ATP: **ATP synthase**.

✓ *Quick check 4*

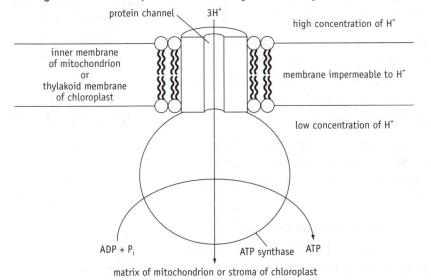

All enzymes have the ending –ase. A 'synthase' synthesises a product, in this case ATP.

Remind yourself of enzyme active sites from your AS course.

The head of a molecule of ATP synthase has three catalytic sites, shown in transverse section opposite, which pass sequentially through three phases of binding ADP and P_i, forming tightly bound ATP and finally releasing it. Part of the molecule γ rotates. This produces the structural changes in the binding sites that allow catalysis.

Some ATP is synthesised by a different method. In two stages of respiration, **glycolysis** and the **Krebs cycle**, some ATP is synthesised *directly* from the energy released by reorganising chemical bonds (see pages 6 and 8). This is called **substrate-level phosphorylation**.

✓ *Quick check 5*

? ## Quick check questions

1. State which of the following combinations of molecules gives a **nucleotide**: (i) ribose + adenine; (ii) ribose + adenine + phosphate.
2. Distinguish between an energy currency molecule and an energy storage molecule.
3. Identify three different types of work performed by one of your muscle cells.
4. Describe briefly how ATP is produced using the potential energy of a proton gradient.
5. Suggest why structural changes in the binding site of an enzyme allow catalysis to occur.

Questions 1 and 5 are synoptic.

Respiration

Respiration is the transfer of chemical potential energy from organic molecules to ATP in living cells. Most cells use carbohydrate, usually glucose, as their fuel. Some cells, e.g. nerve cells, can *only* use glucose as their respiratory substrate but others can use fatty acids, glycerol and amino acids.

Respiration of glucose has four main stages:

- **glycolysis** in the cytoplasm (cytosol) of the cell;
- the **link reaction** in the matrix of a mitochondrion;
- the **Krebs cycle** in the matrix of a mitochondrion;
- **oxidative phosphorylation** on the inner mitochondrial membrane.

> The material between the plasma membrane and the nucleus of a cell is known as **cytoplasm**. This includes the various organelles. **Cytosol** is the matrix in which organelles are suspended.

✓ *Quick check 1*

Glycolysis

The process of respiration begins with the splitting of glucose in the cytoplasm (cytosol) of a cell. After many steps, the 6-carbon (hexose) glucose is converted into 2 molecules of **pyruvate**, each with 3 carbon atoms. Energy from ATP is needed in the first two steps, called **phosphorylation**, but energy that can be used to make ATP is released in the later stages. Glycolysis is summarised below.

> 'Lysis' means 'splitting' or 'breakdown'.

✓ *Quick check 2*

- A molecule of **glucose** is phosphorylated, using two molecules of ATP, to give **hexose bisphosphate**. This phosphorylation converts an energy-rich but unreactive molecule into one that is much more reactive and whose chemical potential energy can be trapped more efficiently.

✓ *Quick check 3*

- The hexose bisphosphate is split into two **triose phosphate** molecules.
- Hydrogen atoms and phosphate groups are removed from the triose phosphate, which is oxidised to two molecules of **pyruvate** (pyruvic acid).

During the process of glycolysis there is a small yield of ATP by substrate-level phosphorylation (see page 5), giving a **net** gain of two ATP molecules per glucose molecule used. Hydrogens removed during glycolysis are transferred to the hydrogen carrier molecule **nicotinamide adenine dinucleotide (NAD)** (shown below) to give reduced NAD (NAD$_{red}$).

NAD is made of two nucleotides linked together. Both nucleotides contain ribose sugar. One nucleotide contains the nitrogenous base adenine; the other has a nicotinamide ring that is the part of the molecule that can accept a hydrogen ion and two electrons. NAD that has accepted these is reduced and is written NAD$_{red}$. NAD is present in cells in small quantities and is continually recycled.

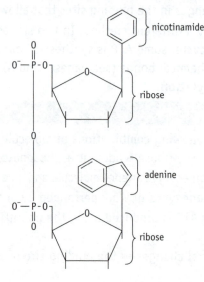

> The reduction of NAD is represented fully as: NAD$^+$ + 2H $\rightarrow$ NADH + H$^+$.

> You do not need to memorise the structure of NAD.

✓ *Quick check 4*

The main stages of glycolysis are shown in the figure.

✓ *Quick check 5*

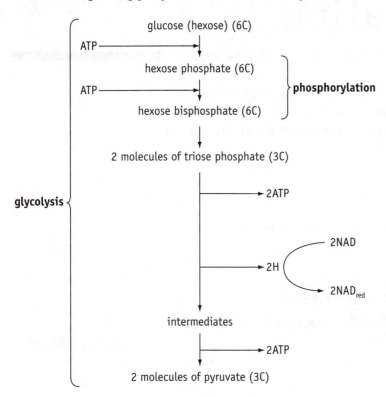

Starting point of glycolysis: one molecule of glucose.

Products of glycolysis: two molecules of pyruvate, a net gain of two molecules of ATP, two molecules of NAD$_{red}$.

? Quick check questions

1 State where glycolysis occurs in a living cell.
2 What is meant by a hexose sugar?
3 Explain why phosphorylation of glucose is necessary at the start of respiration.
4 Describe the role of NAD in glycolysis.
5 Draw up a balance sheet of ATP use and production in glycolysis.

Aerobic respiration

The pyruvate formed in glycolysis is still energy-rich. It passes next to the **link reaction**. This and all subsequent stages of respiration occur inside a mitochondrion and can only occur in the presence of free oxygen. Respiration requiring free oxygen is **aerobic**.

The link reaction

In the link reaction, pyruvate enters the matrix of a mitochondrion and is:

- decarboxylated (carbon dioxide removed);
- dehydrogenated (hydrogen removed);
- combined with coenzyme A to give **acetylcoenzyme A (ACoA)**.

$$\text{pyruvate (3C) + coenzyme A} \xrightarrow[\substack{\text{NAD} \quad \text{NAD}_{red}}]{\text{carbon dioxide}} \text{acetyl (2C) coenzyme A}$$

✓ Quick check 1

Coenzyme A consists of:

- adenine;
- ribose (together with adenine making a nucleoside);
- pantothenic acid (a vitamin).

Coenzyme A transfers an acetyl group (with 2 carbon atoms) from pyruvate into the Krebs cycle and hence plays a central role in respiration. It is present in small quantities in a cell and is recycled.

▶ NAD and FAD are also coenzymes. NAD and FAD are hydrogen carriers while coenzyme A carries an acetyl group.

Krebs cycle

The **Krebs cycle**, or **citric acid cycle** or **tricarboxylic acid (TCA) cycle**, also occurs in the matrix of the mitochondrion. The Krebs cycle is summarised below.

- An acetyl group with 2 carbon atoms from ACoA is combined with a 4-carbon compound (oxaloacetate) to give a 6-carbon compound (citrate or citric acid). Coenzyme A is reformed.

✓ Quick check 2

- Citrate is then converted back to oxaloacetate in a series of small steps involving decarboxylation and dehydrogenation.
- The carbon dioxide removed is given off as a waste product.
- The hydrogens removed are accepted by NAD or the related **flavin adenine dinucleotide (FAD)**. One FAD and two NAD molecules are reduced to FAD_{red} and NAD_{red}, respectively, for each turn of the cycle. The main role of the Krebs cycle in respiration is to generate a 'pool' of reduced hydrogen carriers to pass on to the next stage.

✓ Quick check 3

▶ You do *not* need to know the details of the Krebs cycle, only that citrate is converted to oxaloacetate in a series of small steps.

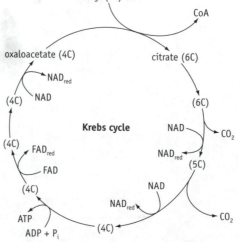

- The regenerated oxaloacetate can combine with another ACoA.
- One molecule of ATP is made directly by substrate-level phosphorylation (see page 5), for each ACoA entering the cycle. Hence two molecules of ATP are made per glucose molecule entering glycolysis.

Electron transport chain and oxidative phosphorylation

These processes take place on and within the inner membrane of the mitochondrion: the **cristae**.

✓ *Quick check 4*

The energy for phosphorylating ADP to ATP comes from the activity of the **electron transport chain (ETC)**, shown opposite. Hydrogens from FAD_{red} and NAD_{red} first pass to **hydrogen carriers** in the inner membrane and are then split into

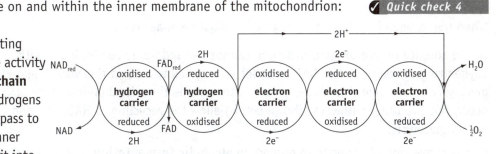

hydrogen ions (H^+) and electrons (e^-). The electrons pass along a series of **electron carriers**, each of which is at a lower energy level than its predecessor. The hydrogen ions remain in solution.

The final electron acceptor is oxygen. When oxygen accepts an electron, a hydrogen ion is drawn from solution to reduce the oxygen to water. Hence, ETC and oxidative phosphorylation need free oxygen to occur.

The transfer of electrons along the series of electron carriers makes energy available for the synthesis of ATP from ADP and P_i by creating a proton gradient across the inner mitochondrial membrane (see page 5).

Potentially, three molecules of ATP can be made from each NAD_{red} entering the ETC and two molecules of ATP from each FAD_{red}, provided that ADP and P_i are available *inside* the mitochondrion. Energy used in transporting ADP into, and ATP out of, the mitochondrion results in the actual yield being about two and a half molecules of ATP from each NAD_{red} and one and a half molecules from each FAD_{red}.

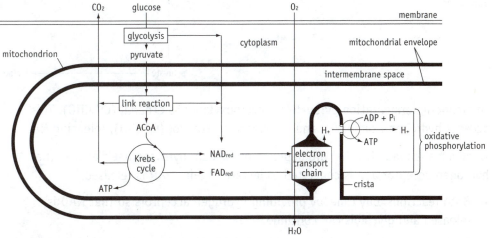

The sequence of events in respiration and their sites in the cell are shown above.

✓ *Quick check 5*

❓ Quick check questions

1. Summarise the events of the link reaction.
2. What is the role of coenzyme A in aerobic respiration?
3. Calculate the number of reduced NAD and reduced FAD molecules produced for each molecule of glucose in aerobic respiration.
4. State where each of the following processes occurs in a living cell: (i) link reaction; (ii) Krebs cycle; (iii) oxidative phosphorylation.
5. Calculate the number of ATP molecules that are formed for every glucose molecule that is respired aerobically.

Anaerobic respiration; respiratory substrates

When free oxygen is not present, respiration must be **anaerobic**.

In the absence of free oxygen, hydrogen cannot be used up by combining it with oxygen to give water, and so NAD_{red} cannot be recycled to NAD in this way to allow glycolysis to continue. The stages of respiration inside the mitochondrion cannot occur, but two other pathways (shown below) allow the recycling of NAD_{red} formed during glycolysis:

● the conversion of pyruvate to ethanol in **alcoholic fermentation**, e.g. by yeast;

● the conversion of pyruvate to lactate (lactic acid), e.g. by mammalian muscle.

Lactic acid ($CH_3CHOHCOOH$) occurs in the body as the anion, lactate.

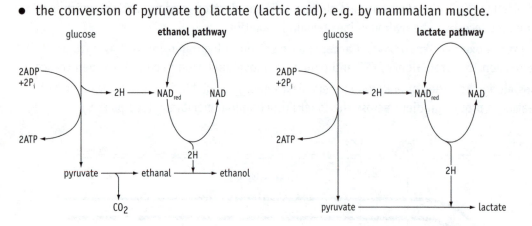

In alcoholic fermentation pyruvate is decarboxylated to ethanal (CH_3CHO). This accepts hydrogen from NAD_{red} and is reduced to ethanol (C_2H_5OH), releasing NAD.

In mammalian muscles that are deprived of oxygen, pyruvate itself acts as the hydrogen acceptor and is converted to lactate. Again, NAD is released.

✓ *Quick check 1, 2*

● Both reactions 'buy time' by providing hydrogen acceptors so that NAD is released and glycolyis can continue.

● Both pathways are inefficient and wasteful in that the products (ethanol or lactate) have chemical bond energy that is untapped.

● The ethanol or lactate produced is toxic and restricts the use of the pathways.

● While the lactate pathway is reversible (by the Cori cycle) in the mammalian liver, the ethanol pathway is irreversible.

● There is a net gain of only two ATPs per glucose (from glycolysis) during anaerobic respiration.

✓ *Quick check 3*

The Cori cycle

In mammals, lactate produced during strenuous muscle activity is taken up from blood plasma by the liver, where it is converted to pyruvate and thence to glucose or glycogen. The cycle is shown opposite.

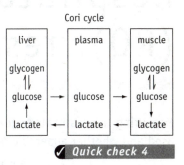

The Cori cycle serves two purposes:

- it 'rescues' lactate and prevents the wasteful loss of some of its chemical bond energy;
- it prevents a potentially disastrous fall in plasma pH.

✓ *Quick check 4*

Respiratory substrates

Glucose is an essential fuel for some cells, e.g. brain cells, red blood cells and lymphocytes, but some cells, e.g. liver cells, also oxidise lipids and excess amino acids (see opposite). It is the fatty acid components of lipids that are important; carbon atoms are detached in pairs as ACoA and fed into the Krebs cycle (see page 8). Amino acids are deaminated and their carbon–hydrogen skeletons converted into pyruvate or into ACoA. The energy values of these different substrates are not the same.

Energy values of respiratory substrates

Most of the energy released in respiration comes from the oxidation of hydrogen to water. Hence, the more hydrogens in the structure of a molecule, the greater the energy value. Fatty acids have more hydrogens per molecule than carbohydrates, so lipids have a greater energy value per unit mass. Energy values in kJ g^{-1} are determined by burning a known mass of the substance in oxygen in a **bomb calorimeter**. Typical energy values are:

respiratory substrate	energy value (kJ g^{-1})
carbohydrate	16
lipid	39
protein	17

💡 Look up the structures of fatty acids and amino acids.

✓ *Quick check 5*

? *Quick check questions*

1 Name one cell in which (i) alcoholic fermentation occurs; (ii) the lactate pathway occurs.

2 Explain how reduced NAD from glycolysis is dehydrogenated in the absence of free oxygen.

3 List the disadvantages of anaerobic respiration in comparison with aerobic respiration.

4 Describe the role of the Cori cycle in mammals.

5 Explain why lipids have a greater energy value than carbohydrates.

Respiratory quotient (RQ); respirometers

Measuring the **respiratory quotient (RQ)** shows what respiratory substrate is being used and whether anaerobic respiration is occurring.

$$C_6H_{12}O_6 + 6O_2 \rightarrow 6CO_2 + 6H_2O$$

This equation shows that the number of molecules, and hence the volumes of oxygen used and carbon dioxide produced, in **aerobic** respiration of glucose are the same. But when different substrates are used in respiration, the **ratios** of oxygen used and carbon dioxide given off differ.

respiratory quotient (RQ) = $\dfrac{\text{volume of carbon dioxide given out in unit time}}{\text{volume of oxygen absorbed in unit time}}$

or, from the equation, **RQ** = $\dfrac{\text{moles or molecules of carbon dioxide given out}}{\text{moles or molecules of oxygen absorbed}}$

so, the RQ for the aerobic respiration of glucose = $\dfrac{\text{volume } CO_2}{\text{volume } O_2} = \dfrac{6}{6} = \textbf{1.0}$

Suppose that the fatty acid oleic acid (as in olive oil) is respired aerobically:

$$C_{18}H_{34}O_2 + 25.5\ O_2 \rightarrow 18\ CO_2 + 17\ H_2O$$

the RQ for the aerobic respiration of oleic acid = $\dfrac{\text{volume } CO_2}{\text{volume } O_2} = \dfrac{18}{25.5} = \textbf{0.7}$

✔ *Quick check 1*

Make sure you can define RQ as the *ratio* of the volume (or number of molecules) of carbon dioxide given out to that of oxygen absorbed.

✔ *Quick check 2*

Typical RQs for aerobic respiration of different respiratory substrates are:

respiratory substrate	respiratory quotient (RQ)
carbohydrate	1.0
lipid	0.7
protein	0.9

Now suppose that respiration is **anaerobic**. Glucose undergoes alcoholic fermentation in a yeast cell:

$$C_6H_{12}O_6 \rightarrow 2C_2H_5OH + 2CO_2$$

the RQ for the alcoholic fermentation of glucose = $\dfrac{\text{volume } CO_2}{\text{volume } O_2} = \dfrac{2}{0} = \infty$

In practice, some aerobic respiration will happen at the same time as anaerobic respiration, so a small volume of oxygen will be used, making the RQ <2. High RQ values indicate that anaerobic respiration is occurring.

No RQ can be calculated for muscle cells using the lactate pathway in anaerobic respiration, because no carbon dioxide is given off:

$$C_6H_{12}O_6 \rightarrow 2\ \text{lactate } (CH_3CHOHCO_2^-)$$

Respirometers

The rate of oxygen uptake in small terrestrial invertebrates or in germinating seeds can be measured in a **respirometer**, as shown below. In respiration, oxygen is taken in and carbon dioxide given off. To ensure that the respirometer shows **only** oxygen consumption, the carbon dioxide produced must be absorbed by a suitable chemical such as soda-lime or a concentrated solution of sodium or potassium hydroxide. The decrease in volume of the air surrounding the organisms then results from their oxygen consumption. Oxygen consumption in unit time is measured by recording the movement of a drop of coloured water or oil in the capillary tube.

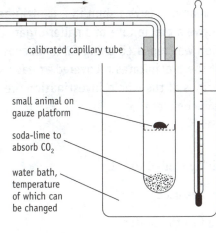

drop of coloured water or oil moves at a rate determined by rate of respiration

calibrated capillary tube

small animal on gauze platform

soda-lime to absorb CO_2

water bath, temperature of which can be changed

✓ *Quick check 3*

- The temperature of the surroundings should be kept constant with a thermostatically controlled water bath whilst readings are taken, as the enclosed volume of air will expand or contract as the temperature changes.
- The apparatus may be used to investigate the effect of temperature on respiration by taking sets of readings at different temperatures.

✓ *Quick check 4*

Effect of temperature on respiration

Respiration is dependent on enzyme activity and so, as the temperature rises, so does the rate of respiration up to the optimum temperature of the enzymes concerned. The rate approximately doubles for a 10°C rise in temperature. The temperature coefficient Q_{10} = 2.

▶ Review the effect of temperature on enzyme action from your AS course.

$$Q_{10} = \frac{\text{rate of reaction at a temperature } t + 10°C}{\text{rate of reaction at a temperature } t}$$

✓ *Quick check 5*

? *Quick check questions*

1 Define the term respiratory quotient.

2 Calculate the respiratory quotient for the aerobic respiration of the following lipid: $2C_{51}H_{98}O_6 + 145 \, O_2 \rightarrow 102 \, CO_2 + 98 \, H_2O$

3 Explain how it is possible to measure oxygen consumption in a respirometer, when the organism is both consuming oxygen and producing carbon dioxide.

4 Why is it necessary to keep the temperature of the surroundings constant and to compensate for atmospheric pressure changes when taking readings with a respirometer?

5 What is meant by the temperature coefficient Q_{10}?

Exam-style questions

1 Liver from a freshly killed mammal was quickly chilled to 2°C. It was macerated with a blender in cold sucrose buffer solution at pH 6.8. The resulting material was passed through a fine gauze to remove cells that had not been broken open by the blender.

The resulting suspension of organelles was spun in a centrifuge at a low speed for a short time. Some material formed a precipitate that was carefully separated from the remaining suspension. This was spun in a centrifuge at a faster speed and for a longer time to give a precipitate of smaller organelles with lower density. The centrifugation was repeated using faster speeds and longer times. The biological activity of the precipitates recovered at each stage was investigated. The conclusions of the whole investigation are summarised in the table below. The table is incomplete.

   a Suggest why:

      i the liver was quickly chilled to 2°C;

      ii a buffer solution was used;

      iii sucrose was added to the buffer.   [3]

cell fraction	organelles in precipitate	biological activity
A: precipitate obtained after slow speed, short time		RNA synthesis
B: precipitate obtained after moderate speed, intermediate time		oxidative phosphorylation
C: precipitate obtained after fastest speed, longest time		protein synthesis
D: solution remaining after centrifugation at fast speed for long time	no organelles	

   b With reference to the table:

      i name the organelle that would be expected in large numbers in cell fractions A, B, C;   [3]

      ii explain why one biological activity expected for cell fraction D would be glycolysis.   [1]

The biological activity of cell organelles can be affected by many chemical factors.

   c State and explain the effect of each of the following on a cell fraction containing many mitochondria:

    i increasing the temperature from 5°C to 15°C;

    ii adding glucose;

    iii adding pyruvate;

    iv adding ADP;

    v adding NAD.   [10]

2 Ethanol is a major constituent of wine and many other drinks. It is produced by yeast cells.

  a Describe how ethanol is produced in a yeast cell.   [3]

Ethanol is transported by the blood from the stomach to the liver. Liver cells contain an enzyme, alcohol dehydrogenase, that oxidises ethanol, using NAD as a coenzyme.

$$CH_3CH_2OH + NAD \quad \xrightarrow{\text{alcohol dehydrogenase}} \quad CH_3CHO + NAD_{red}$$
  ethanol   ethanal

  b **i** Suggest **two** reasons why ethanol is very rapidly absorbed from the stomach into the blood.   [2]

      ii Describe how the reduced NAD resulting from the action of this enzyme might be used in liver cell metabolism.   [2]

3 Batches of germinating seeds from five different species of plant were each placed in a respirometer. The temperature was kept constant and readings were taken when the respirometer contained a carbon dioxide absorber (soda-lime) and also when no soda-lime was present. From the readings, the respiratory quotient (**RQ**) was calculated for each species. The findings of the investigation are summarised in the table.

> RQ is defined in a slightly different way from the definition on page 12. Think carefully about the meaning of rate.

RQ = rate at which carbon dioxide is produced ÷ the rate at which oxygen is absorbed:

species of plant	respiratory quotient (RQ)
A	0.97
B	1.01
C	0.69
D	0.72
E	0.98

With reference to the table:

  a **i** explain why the metabolic rates of the five species **cannot** be compared using only the data from the table.   [1]

Species C produced carbon dioxide at the rate of 12.0 mm^3 min^{-1}:

> Show your working and don't forget units.

      ii calculate the rate of oxygen consumed by species C;   [2]

      iii state and explain which of the five species are likely to store mainly lipid in their seeds.   [2]

> There are 4 marks available so explain in detail and give more than one or two reasons.

  b Explain why the investigation was carried out at a constant temperature.   [4]

Photosynthesis: light-dependent reactions

Photosynthesis is the trapping or fixation of carbon dioxide followed by its reduction to carbohydrate, using hydrogen from water. The necessary energy comes from absorbed light energy.

$$nCO_2 + nH_2O \xrightarrow[\text{chlorophyll}]{\text{light energy}} (CH_2O)n + nO_2$$

Two sets of reactions are involved:

- **light-dependent** reactions in the chloroplast lamellae;
- **light-independent** reactions in the chloroplast stroma.

In the **light-dependent** stage of photosynthesis:

- light energy is trapped;
- ATP is synthesised in **photophosphorylation**;
- water is split by **photolysis** to give hydrogen ions (H^+) that reduce the hydrogen carrier molecule **nicotinamide adenine dinucleotide phosphate (NADP)**.

Oxygen is a waste product of photolysis.

Light energy is trapped in the chloroplast lamellae by photosynthetic pigments that are either:

- chlorophylls, e.g. chlorophylls a and b;
- or carotenoids, e.g. β carotene and xanthophyll.

Different pigments absorb different wavelengths of light, and the absorbance by a pigment can be plotted as its **absorption spectrum**. The effectiveness of the different wavelengths of light in promoting photosynthesis can be shown as an **action spectrum**, as shown below.

> ▶ Do not call the light-independent reactions 'the dark reactions'. They do not need darkness; they simply do not require light.

✓ *Quick check 1*

> ▶ NADP is a slightly different form of NAD (see page 6), with a phosphate group attached to one of the ribose sugars.

✓ *Quick check 2*

✓ *Quick check 3*

> ▶ Note that the absorption peaks for the chlorophylls are in the violet-blue and red regions of the spectrum.

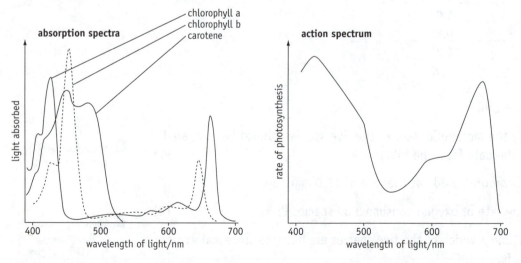

The photosynthetic pigments are arranged as **photosystems** in the chloroplast lamellae. A photosystem is a **light-harvesting cluster** of pigments and each consists of:

- a central **primary pigment** molecule (P), which is one of two forms of chlorophyll a with an absorption peak at either 700 nm **(P700)** in **photosystem I** or 680 nm **(P680)** in **photosystem II**;

- surrounding **accessory pigments**, which are other forms of chlorophyll a, chlorophyll b and the carotenoids. The accessory pigments absorb light and pass energy to the primary pigment.

Cyclic photophosphorylation

When a photon of light is absorbed by photosystem I, an electron in the primary pigment (P700) is excited to a higher energy level and is emitted from the chlorophyll molecule. It is captured by an electron acceptor and passed back to the pigment by a chain of electron acceptors. The energy released allows ATP synthesis from ADP and inorganic phosphate (P_i).

Non-cyclic photophosphorylation

Both photosystems are needed for non-cyclic photophosphorylation. Light is absorbed by both photosystems and electrons emitted from both primary pigments (P700 and P680). Electrons are absorbed by electron acceptors and passed along chains of electron carriers in the **Z scheme**. In this:

- P700 in photosystem I absorbs electrons emitted by photosystem II;

- P680 in photosystem II absorbs electrons from the photolysis of water by enzymes in photosystem II;

$$2H_2O \longrightarrow 4H^+ + 4e^- + O_2$$
$$2NADP \searrow 2NADP_{red}$$

- electrons are used in the synthesis of ATP;

- oxygen is a waste product.

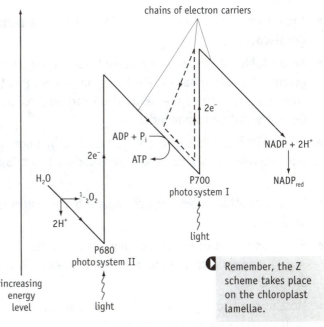

Key — flow of electrons in non-cyclic photophosphorylation
- - - - - - flow of electrons in cyclic photophosphorylation

Remember, the Z scheme takes place on the chloroplast lamellae.

✓ Quick check 4

Comparing cyclic and non-cyclic photophosphorylation:

	cyclic photophosphorylation	non-cyclic photophosphorylation
photosystem	I	I and II
input(s)	light, ADP, P_i	light, water, NADP, ADP, P_i
product(s)	ATP	ATP, $NADP_{red}$, oxygen

✓ Quick check 5

? Quick check questions

1 State precisely where in a chloroplast the light-dependent and the light-independent reactions occur.

2 List the products of the light-dependent reaction of photosynthesis.

3 Distinguish between an absorption spectrum and an action spectrum.

4 Describe the different roles of photosystems I and II.

5 Distinguish between cyclic and non-cyclic photophosphorylation.

Photosynthesis: light-independent reactions

In the **light-independent reactions**, carbon dioxide is first fixed and then reduced to carbohydrate, using the ATP and $NADP_{red}$ produced in the light-dependent reactions. ATP supplies the necessary energy and $NADP_{red}$ the hydrogen atoms needed for the reduction. The reactions involved are cyclic: the **Calvin cycle**.

The main steps of the Calvin cycle are shown below.

- Carbon dioxide, which has diffused into the stroma of a chloroplast, is combined with an acceptor molecule that is a 5-carbon sugar, **ribulose bisphosphate (RuBP)**.

- The enzyme needed for this reaction is **ribulose bisphosphate carboxylase (rubisco)**.

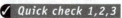 Rubisco is thought to be the most abundant enzyme in the world.

- An unstable intermediate molecule breaks down into two molecules of 3-carbon **glycerate 3-phosphate (GP)**. Plants in which GP is the first detectable compound to include the newly fixed carbon dioxide are called C_3 plants (because GP is a 3-carbon compound).

- Using ATP from photophosphorylation and hydrogen from $NADP_{red}$, both from the light-dependent reactions, the GP is reduced to **triose phosphate**, a 3-carbon sugar.

- Triose phosphate does not accumulate but is immediately converted to other products.

- About one-sixth of the triose phosphate is converted to **hexose** (6-carbon sugar), and then to a range of carbohydrates or is used to synthesise lipids and amino acids.

- Most of the triose phosphate is needed to regenerate RuBP, completing the cycle. This regeneration requires ATP.

✓ *Quick check 1,2,3*

How is RuBP, a 5-carbon sugar, produced from 3-carbon sugars? This process involves a complex cycle including 3-, 4-, 5-, 6- and 7-carbon sugars. The overall result is:

$$5C_3 \rightarrow 3C_5$$

Of 12 molecules of triose phosphate produced in the Calvin cycle:

- 10 are converted to 6 molecules of RuBP (the additional phosphate groups needed come from ATP);

- 2 are *either* converted to 1 molecule of hexose, which may be used to synthesise starch for storage, cellulose for cell wall construction or be converted into other sugars;

- *or* are used to form 2-carbon acetyl groups, which are converted to fatty acids (via acetylcoenzyme A);

- *or* are used to provide the elements carbon, hydrogen and oxygen for amino acid synthesis – nitrogen and sulphur are also needed if amino acids, and hence proteins, are to be made, plants take up these elements as inorganic salts (nitrates and sulphates).

> ◗ Remind yourself of the structures of carbohydrates, fatty acids and amino acids.

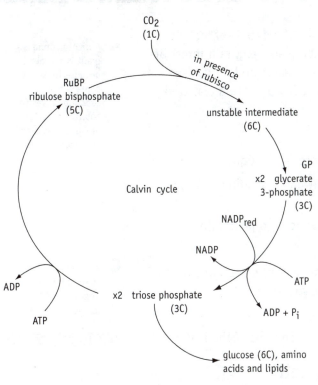

✓ *Quick check 4, 5*

? Quick check questions

1 What is meant by the term carbon dioxide acceptor?

2 State the reaction catalysed by the enzyme rubisco.

3 State the roles of ATP and $NADP_{red}$ in the Calvin cycle.

4 List the possible fates of triose phosphate formed in the Calvin cycle.

5 Explain what may happen to the carbon atoms of 12 triose phosphate molecules formed during the Calvin cycle.

Leaf structure and function

The leaf of a plant is an **organ** adapted for photosynthesis. The details of leaf structure vary greatly and relate to the environment of the plant, e.g. rain forest or desert. A leaf of a **dicotyledonous** plant has a broad, flattened, thin **lamina** (see the figure) and a **midrib** giving rise to a network of veins. It may have a leaf stalk or **petiole**.

A typical leaf:

- is held more or less horizontally to maximise the absorption of light;
- is arranged on the plant in a **leaf mosaic** to avoid unwanted shading by other leaves;
- has a transparent cuticle and upper epidermis to allow light to penetrate;
- is thin and porous so that gases can diffuse quickly;
- has xylem for support and to transport water for photosynthesis and minerals for making various molecules from the products of photosynthesis;
- has phloem to remove the products of photosynthesis;
- has stomata, which when open allow gases to diffuse in and out of the leaf.

> ◖ A dicotyledonous plant is so-called because its seeds contain two seedling leaves (or cotyledons). Most broad-leaved plants are dicotyledons.

> ◖ Revise the structure of xerophyte leaves from your AS course.

> ✓ *Quick check 1*

The structure of a palisade mesophyll cell

The **palisade mesophyll cells** are the main site of photosynthesis. These cells:

- are cylindrical to give air spaces around them;
- are long to avoid too many light-absorbing cell walls;
- have thin cell walls to allow easy diffusion of gases;
- have a thin layer of cytoplasm surrounding a large vacuole, thereby ensuring that the chloroplasts are close to the outside of the cell;
- have many chloroplasts that can be moved to the top of the cell at low light intensities to absorb the available light or to the sides of the cell at higher light intensities to avoid damage to their pigments and to allow light through to the spongy mesophyll below.

> ✓ *Quick check 2*

Chloroplast structure

A **chloroplast** is the eukaryotic organelle adapted for photosynthesis. In dicotyledons, chloroplasts are biconvex discs about 3–10 μm in diameter (see opposite). Each has:

- a surrounding **envelope** of inner and outer phospholipid membranes;
- a liquid matrix or **stroma** where the Calvin cycle takes place and which contains the enzymes needed for the light-independent reactions, including rubisco (see page 18);
- a series of flattened fluid-filled membranous sacs or **thylakoids**, which in places form stacks called **grana** connected by **lamellae**;
- small (70S) ribosomes, as in bacteria;

- DNA circles (loops);
- lipid droplets and starch grains.

The lamellae, and especially the grana, provide a very large surface area for holding the light-absorbing photosynthetic pigments (see page 16) and the electron carriers and enzymes needed in the light-dependent reactions of photosynthesis.

✔ *Quick check 3, 4*

▶ Compare the structure of a chloroplast with that of a prokaryotic cell from your AS course.

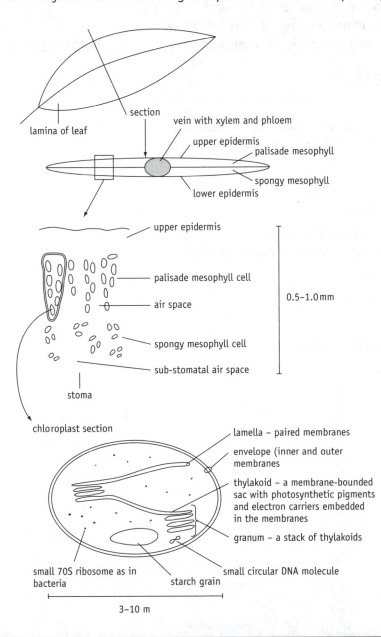

Quick check questions

1 Explain how a 'typical' leaf is adapted for photosynthesis.

2 In what ways is a palisade mesophyll cell well adapted for photosynthesis?

3 State where (i) photophosphorylation and (ii) the Calvin cycle occur in a chloroplast.

4 Draw and label a section of a chloroplast to show its structure.

Limiting factors

When several factors influence the rate of a reaction, the factor that is least favourable will determine the rate at which a reaction can proceed. This is the **law of limiting** factors and the factor that is most unfavourable is the **limiting factor**.

✓ *Quick check 1*

Limiting factors in photosynthesis

To study the factors or variables that limit photosynthesis, each must be changed while all other possible variables remain constant. The factors or variables that may limit photosynthesis are:

- light intensity;
- concentration of carbon dioxide outside the plant;
- temperature (this affects the activity of enzymes catalysing the light-independent reactions and also the rate of diffusion of carbon dioxide);
- degree to which the stomata are open or closed;
- water supply (which influences stomatal closure);
- structure of the leaf;
- availability of chlorophyll, carrier molecules and enzymes.

When the **light intensity** is varied and all other factors remain constant, the rate of photosynthesis increases linearly with increased light intensity over a range of low intensities. The light intensity is the limiting factor. At high light intensities, increasing the intensity has little or no effect (see the graph opposite). A factor *other* than light intensity is limiting the rate of reaction: perhaps the available carbon dioxide.

Notice how the line of the graph becomes horizontal at high light intensities.

As with increasing the light intensity, increasing the concentration of carbon dioxide initially increases the rate of photosynthesis. Over this range of concentrations the **carbon dioxide concentration** is rate limiting. At higher concentrations some other factor is rate limiting.

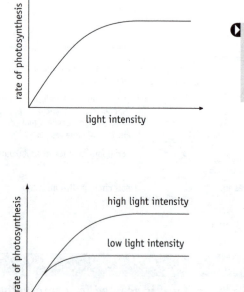

✓ *Quick check 2,3*

At low light intensities, increasing the temperature has little effect on the rate of photosynthesis, because light intensity is then the limiting factor. But at high light intensities, increasing the **temperature** (within limits) increases the rate of photosynthesis, by increasing the rate of the light-independent stages. At higher

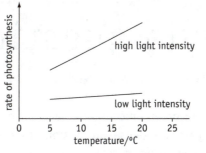

Revise the effect of temperature on enzyme action from your AS course.

temperatures (above 25–30°C) the proteins of the cell begin to be denatured and the rate falls. Below 0°C, the cells freeze and are killed by ice crystals. This is why the graph covers a small temperature range.

✓ *Quick check 4*

Measuring the rate of photosynthesis

The rate of photosynthesis in different conditions can be measured using a **photosynthometer** (see the diagram below). By using this apparatus, it is possible to measure the rate of oxygen production by a water plant (e.g. *Elodea canadensis*) while varying the light intensity (or another factor). Bubbles of oxygen collect in the capillary tube of the apparatus. When a suitable volume of gas has been collected in a known time period, it can be drawn (by a syringe) as a bubble alongside the scale and the length of the bubble measured.

Note that:

- all variables other than the factor being investigated must be kept constant;
- the plant must be left for a time to adjust to new conditions before any measurements are made;
- at least three measurements should be taken in each condition and the mean value found.

barrel of syringe | plastic tube | thermometer | board | plunger of syringe | flared end of capillary tube | capillary tube | ruler | clamp | light | water plant | stand | beaker of water (heat shield) | test tube

Quick check questions

1 Explain what is meant by a limiting factor.

2 List three factors that could limit the rate of photosynthesis at high concentrations of carbon dioxide.

3 Copy the graph shown for carbon dioxide and add a curve to show the rate of photosynthesis at an even lower light intensity.

4 Explain why at low light intensities temperature has little effect on the rate of photosynthesis.

Exam-style questions

1 *Pleurococcus* is a genus of green unicellular algae that grow on damp surfaces such as stones or tree trunks. A dense suspension of *Pleurococcus* cells was prepared in a very dilute solution of potassium hydrogen carbonate ($KHCO_3$). The suspension was taken into a syringe which was attached to a capillary tube, as shown in the figure.

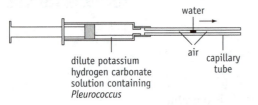

dilute potassium
hydrogen carbonate
solution containing
Pleurococcus

The temperature of the syringe remained constant during the experiment.

When the apparatus was illuminated using a lamp, the small bead of water in the capillary tube moved steadily away from the syringe in the direction shown by the arrow.

a Explain why the bead of water moved in the direction indicated.   [2]

b Describe and explain the movement of the bead of water if the apparatus was in darkness at the same temperature.   [2]

c Describe and explain how the rate of movement of the bead would change:

  i if the syringe were brightly illuminated and the temperature varied between 2°C and 20°C;

  ii if it were illuminated by very dim light and the temperature varied over the same range.   [5]

The hydrogen carbonate ions in the solution could be made radioactive using carbon-14 (^{14}C). Under these conditions, the cells would produce many types of radioactive organic molecules in a short time.

d i Explain how it is possible for many different types of radioactive molecule to be produced.   [2]

  ii Name the first organic substance that would become radioactive in the cells.   [1]

2 a In a typical dicotyledonous plant, stomata are found mainly on the underside of the leaves. Suggest advantages of this arrangement for the plant.   [3]

  b Give **two** adaptations that allow leaves to conserve water on a hot bright day.   [2]

  c Explain how each of the features of a palisade mesophyll cell listed below enables the cell to maximise photosynthetic efficiency:

  i chloroplasts which can move around the cell;

> ◖ This question is partly synoptic. You learnt about leaves in AS Module 2803/01.

    ii long thin shape;

    iii vertical orientation.   [3]

Leaves of many plant species develop differently depending on whether they grow in sunny or in shady conditions. Cuttings from the same *Impatiens* plant were allowed to grow at two different light intensities, in controlled environment chambers, until both cuttings had developed new leaves.

d State **two** variables that would have been kept constant in the controlled environment chambers.   [2]

e Explain why cuttings from a single plant were used rather than seedlings. [1]

Equal masses of leaves from each cutting were placed in apparatus that allowed the volume of oxygen used or produced to be measured. Measurements were made at different light intensities and the results are shown in the graph.

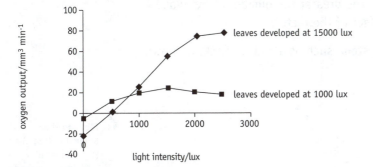

With reference to the graph:

f   **i** explain why both sets of leaves are net absorbers of oxygen at low light intensities;   [2]

    ii suggest why the leaves that developed at a light intensity of 1000 lux both produced and consumed less oxygen than the leaves that developed at 15000 lux.   [4]

3 The fixation of carbon in photosynthesis can be summarised by the equation:

$$3CO_2 + 9ATP + 6NADP_{red} \rightarrow \text{triose phosphate} + 8P_i + 9ADP + 6NADP + 3H_2O$$

With reference to this equation:

a   **i** explain the origin of the ATP and $NADP_{red}$;   [3]

    ii explain why P_i, ADP and NADP do not accumulate in the chloroplasts during carbon fixation.   [1]

b Explain why ribulose bisphosphate (RuBP) does not feature in the summary equation.   [1]

The product of photosynthesis is often glucose or starch.

c Explain briefly how these molecules are produced.   [2]

Populations and Interactions

Most bacteria multiply by **binary fission**, in which one cell divides into two daughter cells, doubling the number of cells every generation. The time taken for a bacterial population to double is called its **generation time**. Under optimal conditions some bacteria can divide as frequently as once every 20–30 minutes. But there are always factors preventing unlimited population growth, such as:

- depletion of nutrients;
- depletion of oxygen;
- accumulation of toxic or acidic waste products.

These **limiting factors** are density-dependent: the greater the number of individuals in the population (density) the greater the effect of the factor.

Growth of a bacterial population in a closed system, such as a culture flask, is shown in the figure below.

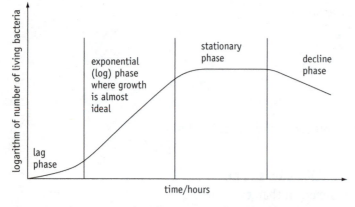

There are four recognisable phases of population growth.

Lag phase: the bacteria adjust to new conditions, synthesising carriers to absorb nutrients or enzymes to digest them. This may involve switching on genes (see *lac* operon, page 48). Growth is slow.

Exponential phase (logarithmic phase): the numbers of bacteria double in each unit of time. The exponential (geometric) growth curve that this gives becomes a straight line when the logarithms ($\log_{10}$) of the numbers of bacteria are plotted against time. There are no restrictions to growth in this phase and growth is rapid.

Stationary phase: limiting factors, such as the failing supply of nutrients and oxygen and the build-up of waste products, gradually have their effect until the death rate equals the division rate and no population growth occurs.

Decline phase (death phase): the death rate is greater than the division rate so the population becomes smaller, even though the decline may be slowed slightly by breakdown of cells releasing nutrients that other cells can use.

Limiting factors in population growth

Any factor that prevents a population from increasing in size is a **limiting factor**. At any given time there is usually just one factor limiting population growth, but

> Plotting the logarithm ($\log_{10}$) of the number of individuals converts the exponential (logarithmic) curve into a straight line.

> You will learn more about bacterial population growth if you study Microbiology and Biotechnology as your optional module.

> ✓ *Quick check 1,2, 3,4*

different limiting factors come into play at different times. The limiting factor may only operate seasonally. Examples of limiting factors for populations of terrestrial flowering plants and animals are listed below.

flowering plants	animals
lack of light	lack of food/prey
lack of water	lack of water
lack of carbon dioxide	lack of oxygen (only likely in aquatic habitats)
lack of nutrient ions	lack of a suitable site for reproduction/egg laying
an inappropriate temperature	an inappropriate temperature
infection by pathogens	infection by pathogens
agent for cross-pollination absent	lack of a mate at very low population density
lack of space (another factor almost always becomes limiting before physical space)	lack of space (many animals defend a territory, ensuring resources for reproduction)

Compare this use of the concept of limiting factors with that in photosynthesis on page 22.

Carrying capacity

The maximum population density of an organism that can be supported permanently in a habitat is called the **carrying capacity** and is determined by one or more of the **density-dependent** limiting factors. Such factors may be:

- **biotic**, caused by other living organisms, e.g. predation, competition or infection by pathogens;

- **abiotic**, involving non-living components of the environment, e.g. water supply or nutrients in the soil.

A density-*dependent* factor has a proportional increase or decrease in its effect as the population density rises or falls. The effect of a density-*independent* factor is not related to population size.

✓ *Quick check 5*

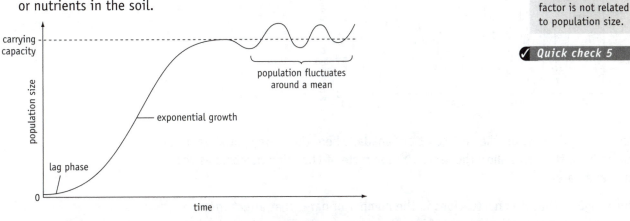

1 Explain why, after a while, growth of a population of bacteria in a closed system ceases to be exponential.

2 Lactobacilli produce lactic acid from the sugar lactose in converting milk to yoghurt. List the density-dependent limiting factors that may affect population growth of these bacteria in a closed system.

3 Predict the shape of a growth curve for a population of bacteria in an open system in which materials can enter and leave.

4 Explain what is meant by a density-dependent limiting factor.

5 Define the term carrying capacity.

Predator–prey interactions

Annual records of the numbers of Canadian lynx and snowshoe hares trapped for the Hudson's Bay Company in Canada have been kept since the 1820s and provide evidence of a 10-year cycle in the numbers of both animals (see the figure below). Both lynx and hare were trapped for their fur. Trappers found that at a peak of the cycle thousands of lynx could be trapped in a season, whilst in a trough, 5 years later, only a few hundred would be trapped.

A possible explanation of this cycle is as follows.

- A small population of snowshoe hares cannot support many lynx, so the lynx population crashes.

- But now there are few lynx to prey on the snowshoe hares, so the hare population starts to rise.

- This in turn supports more lynx, so the lynx population rises.

- But the large numbers of lynx eating snowshoe hares causes the hare population to crash and so the cycle begins again.

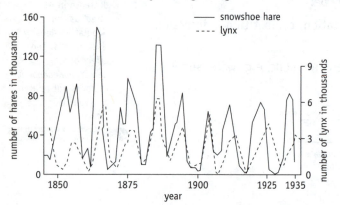

> Look, in a graph such as this, for *trends* and *times* of peaks.

However, on islands off the east coast of Canada, where there are snowshoe hares but no lynx, the hares show the same 10-year cycle of changing numbers as the mainland hares.

This suggests that it is fluctuations in the number of hares that affect the lynx population, but that the numbers of lynx do not cause the fluctuations in the hare population. Instead this may result from:

- changes in the availability of food plants;

- changes in the nutritive quality of food plants;

- reduction in breeding success of hares at high population densities.

✓ *Quick check 1,2*

Other small herbivores, for example lemmings, also show regular cycles in their numbers with no obvious involvement of predators.

Competition

Organisms compete for resources or space that are in limited supply. Such competition falls into two categories:

- **intraspecific** competition, between members of the **same** species;
- **interspecific** competition, between members of **different** species.

Intraspecific competition

Density-dependent factors (see page 27) limit population growth, and the **fitter** individuals with adaptations that are better suited to the conditions prevailing will outcompete the less fit. This is one of the causes of **natural selection** (see page 54).

Interspecific competition

Competition between species is rarely measurable in natural populations. The principle of **competitive exclusion** states that, as a result of competition, two species do not occupy the same **niche**. One displaces the other in such a way that each adopts a certain way of life in which it has an advantage. Hence, data on interspecific competition tend to come from laboratory investigations.

- When the protoctists *Paramecium caudatum* and *P. aurelia* are grown together in laboratory culture in conditions in which both species survive when grown alone, *P. caudatum* always becomes extinct.
- Duckweed, *Lemna minor*, floating on the surface of a tank of water outcompetes its relative *L. trisulca*, which grows submerged in water.
- When two species of flour beetle, *Tribolium confusum* and *T. castaneum*, are placed together in conditions in which both thrive when cultured separately, only one species survives. At lower temperatures and humidity *T. confusum* outcompetes *T. castaneum*. At higher temperatures and humidity the reverse occurs.
- In natural UK habitats, the red squirrel is probably in competition with the introduced larger grey squirrel. Red squirrels outcompete grey in conifer woods, but are outcompeted by grey squirrels in woods with less than 75% conifers.

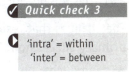

✓ *Quick check 3*

'intra' = within
'inter' = between

A niche is a way of life; a habitat is where an organism lives.

✓ *Quick check 4,5*

? Quick check questions

1. What evidence does the figure opposite provide that the snowshoe hare population controls the lynx population and not vice versa?

2. The numbers of bank voles and wood mice in a wood in the UK show seasonal fluctuations in numbers, peaking in autumn and early winter. Both species are eaten by tawny owls. Predict the effect of the fluctuations in the prey species on owl numbers.

3. Distinguish between **intraspecific** and **interspecific** competition.

4. Explain what is meant by **competitive exclusion**.

5. Suggest explanations for the results of competition between species of *Paramecium* and *Lemna*.

Succession

Succession is a change in the structure and species composition of a community. It occurs because of the changes caused by the presence of established organisms or from some external influence. The immediate changes caused by agricultural activities are not included in this definition.

The changes that result when the starting point is bare uncolonised land are called **primary succession**. This occurs in:

- sand dunes;
- lava and ash from volcanoes;
- landslides;
- land and lakes left by retreating glaciers.

The first colonisers on bare rock are lichens and mosses. Their presence allows some soil to build up, allowing ferns and flowering (vascular) plants to appear. When the bare ground is particulate, e.g. made up of sand or ash particles, colonisation is faster because vascular plants can take root without the lichen and moss stage. Eventually the soil becomes deep enough to support shrubs and then trees. In Europe, woodland is the endpoint of succession on land and is called the **climax community**.

✓ *Quick check 1*

The endpoint of primary succession in a freshwater pond or lake is also woodland. Once the water has been colonised by plants there is an increase in organic matter and the lake gradually fills in.

Primary succession

The effects of named organisms on the process of succession can be studied with examples of primary succession, e.g. sand dune, lake hydrosere or volcanic island. Some of these organisms will be early arrivals (pioneer species) and others characteristic of the climax vegetation.

❶ A **sere** is a particular example of plant succession. A **hydrosere** is a sere beginning in water.

Pioneer species:

- are able to tolerate extreme conditions, e.g. low nutrient levels;
- have very good means of dispersal, usually by wind;
- are not able to compete for resources, e.g. light;
- are not influenced by or dependent upon animal species;
- may be able to fix nitrogen (e.g. legumes) and build up soil nutrients.

❶ List the possible effects each species you have studied could have on succession.

✓ *Quick check 2*

Climax community species:

- have large seeds (with large energy store) so that seedlings can survive low light intensity;
- have a specialised niche, e.g. as an epiphyte;
- are unable to tolerate great fluctuations in water content of soil;
- are strongly influenced by other organisms, e.g. competitors, herbivores, pollinators, seed-dispersal agents and soil micro-organisms.

Sampling a habitat

When studying a habitat you need to know:

- what species are present in the habitat;
- the abundance of each species present;
- the distribution of each species.

Initially organisms must be identified, usually by using an **identification key** or consulting an expert in the field.

Techniques for sampling a habitat

Organisms are distributed unevenly in most environments, so **random sampling** is used to determine the number and abundance of species present. Random sampling also eliminates any bias on the part of the person taking the samples. The area to be sampled is mapped and given a grid of numbered squares of a size appropriate to the sampling technique. The square to be sampled is determined by use of random number tables or random numbers generated by a calculator or computer.

✓ *Quick check 3*

- **Quadrats** are square frames, the size appropriate for the area to be surveyed, within which you can survey what species are present and estimate the abundance of each by: (i) finding the **percentage cover**, the proportion of the quadrat's area occupied by the species; (ii) counting the **number of individuals** present and finding the average number per quadrat; (iii) finding the **species frequency**, the proportion of quadrats with a particular species in them; (iv) using a **subjective scale** such as ACFOR (Abundant, Common, Frequent, Occasional and Rare) or DAFOR (Dominant, Abundant, Frequent, Occasional and Rare). Quadrats are suitable for plants and sessile animals in areas with relatively uniform conditions, where the quadrats must be distributed randomly.

  ▶ Using a subjective scale is quick and avoids counting. Accuracy improves with practice.

- **Point quadrats** are frames through which long pins (usually 10) are lowered vertically. Each species that touches a pin is recorded, together with the total number of times it is touched. The point quadrat must be positioned randomly.

- **Line transects** are lines across a habitat. All the species touching a line are identified and the position of each species that touches the line recorded. Transects are suitable for habitats with gradations in conditions. The starting point of the transect should be random, but the line should be placed across the gradation concerned, e.g. at right angles to the line of a ditch, stream or hedge or from the sea to the high tide zone.

  ▶ You will use these sampling techniques if you study Environmental Biology as your optional module.

- **Belt transects** are quadrats placed sequentially along a line transect, making the transect wider. The quadrats are used as above.

✓ *Quick check 4*

❓ Quick check questions

1 Explain why colonisation is faster on particulate bare ground than on rock.

2 What adaptations would you expect a pioneer species of plant to show?

3 Explain when it is necessary to use random sampling techniques.

4 Choose an appropriate sampling technique for finding the plant species present in (i) a large area of grassland and (ii) a river bank.

Conservation

Conservation of a species, habitat or ecosystem may be needed because it is at risk from human influence or other pressures. It is always a dynamic process because it must respond to change. It may be for the benefit of people or other organisms.

conservation measure	example
protection of species	laws against hunting the blue whale
protection of habitats	laws preventing discharge of wastes into rivers
restoration or reclamation	erecting sea defences, preventing sand being blown off dunes
creating new habitats	digging ponds, planting new woodland
captive breeding of endangered species	breeding programmes for the Hawaian goose
prevention of succession	grazing and/or burning of heaths or grassland
national action	laws preventing destruction of bats and their roosts
international action	CITES treaty preventing trade in products such as rhino horn

Conservation is not the same as preservation. **Preservation** protects species and/or habitats, e.g. by creating a nature reserve, while conservation includes the active management needed to maintain or increase biodiversity. Conservation may be needed to counter the effects of modern farming methods.

✓ *Quick check 1*

Use and effect of N fertilisers

Nitrogen is an essential element for all organisms, needed for proteins and nucleic acids. It is a **macronutrient** in plant growth and is taken up in large quantities in the form of nitrate ions (NO_3^-). High-yielding varieties (HYV) of cereals are dependent on a supply of fixed nitrogen, mainly from inorganic fertilisers, e.g. ammonium nitrate. Excessive use of nitrogen-containing fertilisers has several consequences.

- Nitrate is easily leached from soil, particularly in wet weather, and may end up in watercourses, resulting in **eutrophication** and a high **biochemical oxygen demand (BOD)**. The resulting lack of oxygen kills animal life.
- Nitrate in drinking water may result in the formation of **methaemoglobin**, a stable form of haemoglobin that cannot carry oxygen, in infants and elderly people. Nitrate in drinking water may also contribute to the incidence of stomach cancer.
- Addition of nitrogen-containing fertiliser to soil may reduce the activity of the bacteria of the nitrogen cycle.
- Slow-growing or small species may become locally extinct due to competition.

Strategies for avoiding or minimising the use of artificial fertilisers include:

- crop rotation or intercropping with leguminous plants, which have symbiotic **nitrogen-fixing** bacteria in their root nodules;
- use of organic sources of fixed nitrogen, e.g. animal manure and sewage sludge;
- future genetic modification of nitrogen-fixing bacteria to form associations with non-leguminous plants, especially cereals;
- future addition, by genetic engineering, of the enzyme nitrogenase into non-leguminous plants.

✓ *Quick check 2*

The fixed nitrogen incorporated into grazed pasture or into harvested crop is removed: it is not available for recycling via the nitrogen cycle.

✓ *Quick check 3*

Much nitrate in water does **not** come from inorganic fertiliser but from ploughing up grazing land in which large quantities of organic matter have accumulated.

Look up the nitrogen cycle from your AS course.

Look up gene manipulation from your AS course.

Management of an ecosystem

Many ecosystems provide economically important resources, e.g. aquatic and marine ecosystems yield fish and natural grasslands yield meat and other animal products. **Sustainable management** allows the same area to be exploited indefinitely:

- it does not result in loss of fertility, e.g. by soil erosion;
- the population that is exploited does not become extinct or seriously decline;
- biodiversity is maintained, and the destruction of other species that share the habitat with the exploited species is avoided.

Unsustainable management results in damage to the habitat and/or the depletion of the exploited species to a point where it no longer provides an economic return, as has happened to cod over much of the North Sea.

Sustainable forestry

Timber has great economic value and is used in construction, as fuel and in paper manufacture. Some timber is grown on a small scale on agricultural land, e.g. as part of hedgerows, but most is from woodlands and forests. These may have developed naturally or may be plantations.

There are different ways forests or woodlands can be exploited sustainably.

- **Selective felling**: some mature trees, diseased trees and unwanted species are harvested, leaving other trees to develop and distribute seeds to fill the gaps. This procedure is more expensive than cutting all the trees on the site (**clear felling**).

- **Strip felling**: small patches or strips of forest are completely cleared, leaving other patches untouched to cut many years later after the first areas have regrown. Large areas are not felled at the same time, so loss of species and soil erosion are avoided.

- **Coppicing**: trees are cut down leaving stumps from which new shoots develop. These grow rapidly because there is a well-developed root system. After a few years the shoots are cut and yield poles, but not large logs. Coppicing can be repeated indefinitely. Small strips or patches are cut in different years, producing a high biodiversity because areas of coppice of different age provide a variety of habitats. Use of coppicing had declined because there was little demand for wooden poles. There is new interest in the method because species of willow can produce very large masses of wood in a few years. The wood is unfit for construction work, but excellent for paper manufacture or to burn in power stations to generate electricity.

> Much easily available information involves tropical rain forest. Some of this is relevant but OCR has specified *temperate* forest.

> ✓ *Quick check 4*

> Only some species of tree can be coppiced. Pines and firs do not develop new shoots from cut stumps. Hazel and sweet chestnut can be used.

? Quick check questions

1. Explain the difference between conservation and preservation of species.
2. Explain why HYV cereals need a large supply of fixed nitrogen.
3. What would be an inappropriate time for applying fertiliser?
4. Explain the difference between selective felling and clear felling of woodland.

Exam-style questions

1 Production of zinc, lead and cadmium takes place on a large scale at Avonmouth near Bristol. Cadmium is a toxic heavy metal. It is estimated that about 3.5 tonnes per year of cadmium escape into the atmosphere from the Avonmouth metal smelter. Most of this is deposited onto soil and vegetation within a few kilometres of the smelter.

> This question is synoptic.

    a State an effect of heavy metal ions, such as lead and cadmium, on living cells. [1]

The effect of the Avonmouth pollution was studied by comparing two oak woodlands. Hallen Wood is 3 km from the smelter and Wetmoor Wood is 23 km from the smelter.

The table shows the population density of animals inhabiting the litter and surface layers of soil and the dry mass of litter (dead plant remains) in these woodlands. The types of animal listed are detritivores, feeding mainly on dead and decaying leaves and twigs.

	mean number of individuals per m^2		ratio Wetmoor/Hallen (to nearest whole number)
animals	Hallen Wood	Wetmoor Wood	
earthworms	17	75	4:1
millipedes	8	79	10:1
woodlice	56	171	
mites	129 000	194 000	1:1
springtails	20 800	8 688	
	mean mass of litter/ kg m^{-2}		
mass of litter	14.28	1.35	1:11

    b Complete the table by inserting the appropriate ratio for woodlice and for springtails. [2]

    c With reference to the table:

       i explain the difference in the mass of litter that has accumulated in Hallen Wood compared with Wetmoor Wood; [1]

       ii suggest why the ratio for springtails is different to the ratios for the other types of animal. [2]

The following table shows the concentration of cadmium in the bodies of several organisms collected in Hallen Wood and in the litter from the wood.

	concentration of cadmium/ mg kg^{-1} (dry mass)
litter	23.5
grass (*Holcus lanatus*)	5.0
grass (*Milium effusum*)	3.8
bluebell	14.2
slug (*Arion hortensis*)	56.5
slug (*Arion alter*)	119.3
woodlouse (*Oniscus asellus*)	125.7
woodlouse (*Porcellio scaber*)	41.1
earthworm	56.6

d With reference to the second table:

   i calculate the mean concentrations of cadmium in the living plants and in the animals; [2]

   ii explain why the concentration of cadmium is different in the plants and the animals; [2]

   iii suggest why pairs of similar species of animal, e.g. the slugs *Arion hortensis* and *A. alter*, have different cadmium concentrations. [2]

e Describe briefly how you would measure the abundance of bluebells, or another small plant species, in a community such as woodland. [2]

When seedlings or cuttings are grown in solutions of nutrient salts to which small quantities of cadmium have been added, root growth is poor and they fail to thrive. When this technique was used to compare the populations of the grass *Holcus lanatus* found in Hallen Wood and Wetmoor Wood, the plants from both woods grew equally well in the absence of cadmium. Those from Hallen Wood grew more strongly and with healthier roots than those from Wetmoor Wood when cadmium was present.

f Explain how the population in Hallen Wood has become relatively tolerant of cadmium. [7]

In this question, 1 mark is available for the quality of written communication.

Meiosis, Genetics and Gene Control

Meiosis is the process by which a nucleus divides by two divisions into four nuclei, each containing half the number of chromosomes of the mother cell. The resulting nuclei are **haploid (*n*)**. Meiosis is necessary for sexual reproduction in eukaryotes; without it fertilisation would double the chromosome number every generation.

> You should remind yourself of the details of mitosis before tackling meiosis.

Division I

In **prophase I**, homologous chromosomes **(homologues)**, each consisting of two genetically identical sister chromatids held together by their centromere, form pairs in the process of **synapsis**. Each pair is called a **bivalent**. **Crossing-over** occurs between the homologous chromosomes of a bivalent (see opposite). The chromosomes condense and become more visible, as in mitosis. At the end of prophase I the nuclear envelope breaks down. Centrioles, if present, duplicate and move to opposite poles. A spindle of microtubules forms.

> A bivalent consists of *two* chromosomes and so *four* chromatids.

> A microtubule is a hollow cylinder made of the protein tubulin.

In **metaphase I**, bivalents line up independently on the cell's equator and each homologue attaches to spindle microtubules by its undivided centromere. The orientation of the maternal and paternal chromosomes of each bivalent is random. This **independent assortment** of paternal and maternal chromosomes gives further genetic diversity.

Homologues, each consisting of two chromatids that are no longer genetically identical because of crossing-over, are moved apart to opposite poles by the shortening of the microtubules of the spindle in **anaphase I**.

The chromosomes reach opposite poles of the cell in **telophase I**. After this, new nuclear envelopes form and the cell may divide into two haploid cells.

Division II

In **prophase II**, centrioles, if present, duplicate and move to the poles. A new spindle forms. The nuclear envelopes break down.

The chromosomes are moved to the equator in **metaphase II** and attach to the spindle by their centromeres, which then divide.

Then, in **anaphase II**, sister chromatids are separated.

Finally, in **telophase II**, new nuclear envelopes form and the cells divide.

division I
prophase I
metaphase I
anaphase I
telophase I

division II
prophase II
metaphase II
anaphase II
telophase II

In **meiosis II**, the chromatids of a homologue separate as in a mitotic division. The final result is four nuclei, each with half the chromosome number of the original nucleus and all genetically different from one another.

✓ *Quick check 1, 2*

Genetic variation

Meiosis and fertilisation lead to genetic variation. **Genetic variation** is generated in three main ways.

- **Crossing-over**: during prophase I of meiosis, non-sister chromatids reciprocally swap portions of chromatids, giving new combinations of alleles. In a few organisms, e.g. male *Drosophila* sp., no crossing-over takes place, but in most eukaryotes the number of cross-overs is proportional to the length of the chromosome.

chiasma

chromatids may break and reciprocally exchange pieces

✓ *Quick check 3*

- **Independent assortment**: at metaphase I of meiosis, there is a 50:50 chance which way round a pair of homologous chromosomes (a bivalent) will be placed on the equator of the cell. This gives 2^n possible combinations of paternally derived and maternally derived chromosomes in the daughter cells (where n is the haploid number of chromosomes).

either   or

✓ *Quick check 4*

- Meiosis halves the chromosome number and this is restored during fertilisation by the **random** fusion of two gametes to give a zygote.

Such a comparison is synoptic; it will *not* be examined in Module 2804.

It is useful to compare mitosis and meiosis.

mitosis	meiosis
maintains the chromosome number	halves the chromosome number
has one division cycle	has two division cycles
gives two daughter nuclei	gives four daughter nuclei
does not involve crossing-over and independent assortment	involves crossing-over and independent assortment
gives daughter nuclei that are genetically identical (bar mutation) to one another and to the parent nucleus	gives daughter nuclei that are genetically different from one another and from the parent nucleus

✓ *Quick check 5*

? Quick check questions

1 Describe the essential differences between meiosis I and meiosis II.

2 State the similarities between meiosis II and mitosis.

3 Explain what occurs in a bivalent during crossing-over.

4 In a male *Drosophila*, only independent assortment produces genetic variation in meiosis. The diploid ($2n$) chromosome number is 8. Calculate the number of genetically different spermatozoa that can be produced.

5 Draw out and complete the table below to show the different outcomes of mitotic and meiotic divisions of a cell whose diploid number is 16.

Questions 2 and 5 are synoptic.

	mitosis	meiosis
number of division cycles		
number of daughter cells		
number of chromosomes per nucleus in daughter cells		

Genetics: terminology, monohybrid crosses and codominance

Genetics is the study of genes, their effects and their inheritance. It also encompasses DNA technology, which includes genetic manipulation in, for example, genetic engineering and gene therapy. **Genomics** is the attempt to map and describe all the genes of various organisms.

Genetic terminology

gene	A length of DNA coding for the production, via mRNA, of a polypeptide or coding for the direct production of rRNA or tRNA.
allele (allelomorph)	A variant form of a particular gene. There may be two, several or very many alleles of a gene.
genotype	The alleles present in an individual or cell.
genome	The sum total of genes in an organism.
phenotype	The expression of the alleles of the genotype giving the individual's traits.
homozygote	A genotype in which the two alleles of a gene are identical, e.g. AA or aa.
heterozygote	A genotype in which the two alleles of a gene are different, e.g. Aa.
dominant allele	A dominant allele has the same effect on the phenotype when heterozygous as when homozygous. It is symbolised by a suitable upper case letter, e.g. T for tall.
recessive allele	A recessive allele only affects the phenotype when homozygous. It is symbolised by a lower case letter, e.g. t for short, and must be the same letter as used for the dominant allele.
codominant alleles	Codominant alleles both affect the phenotype. They are symbolised by suitable upper case superscripts to an upper case letter representing the gene, e.g. I^A and I^B for human blood groups A and B.
multiple alleles	Most genes have many alleles; some very many.
locus	The particular position of a gene on its chromosome.
homologous chromosomes (homologues)	Homologues have the same genes in the same sequence (except chromosome mutation) but the alleles of the genes may be different on the paternal and maternal chromosomes. Homologous chromosomes are able to form bivalents during prophase I of meiosis.
sex linkage	Genes that are inherited together on a sex chromosome show sex-linked inheritance, e.g. the X chromosome.
F_1: first filial generation	F_1 should only be used to denote the first generation resulting from crossing two homozygotes. In other circumstances 'offspring 1' should be used.
F_2: second filial generation	F_2 should only be used when crossing two members of the F_1 generation. In other circumstances 'offspring 2' should be used.
test cross	A cross to test whether an individual with a dominant phenotype is homozygous or heterozygous. The individual is crossed with a homozygous recessive individual.

Genetic diagrams: monohybrid crosses

A **genetic diagram** is the accepted way of showing the genotypes and phenotypes of the parents and expected offspring of particular parents. A **monohybrid cross** is concerned with the inheritance of **one** gene; examples are shown below.

The inheritance of red, pink or white flower colour in snapdragons (*Antirrhinum*) is controlled by two alleles that show **codominance**. Allele C^R gives red flowers and allele C^W white flowers. Three genotypes and three phenotypes are possible:

genotype	phenotype
$C^R C^R$	red flowers
$C^R C^W$	pink flowers
$C^W C^W$	white flowers

The symbols for the alleles follow the convention for codominant alleles. Gametes are shown with circles round them.

Consider a cross between a red-flowered and a white-flowered snapdragon:

parental phenotypes	red flowers	white flowers
parental genotypes	$C^R C^R$	$C^W C^W$
genotypes of gametes	all $\boxed{C^R}$	all $\boxed{C^W}$
genotype of offspring (F₁)	all $C^R C^W$	
phenotype of offspring	pink flowers	

Now consider a cross between a pink-flowered and a white-flowered snapdragon:

parental phenotypes	pink flowers	white flowers
parental genotypes	$C^R C^W$	$C^W C^W$
genotypes of gametes	$\boxed{C^R}$ or $\boxed{C^W}$	all $\boxed{C^W}$
	in equal proportions	

Use a **Punnett square** to predict the results of a cross, even in simple examples where it may seem obvious.

genotypes and phenotypes of offspring

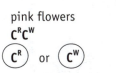

		gametes from white parent C^W
gametes from pink parent	C^R	$C^R C^W$ pink flowers
	C^W	$C^W C^W$ white flowers

All the gametes from the white parent are the same (C^W). Therefore only use C^W once in the Punnett square.

Approximately equal numbers (a ratio of 1:1) of pink- and white-flowered offspring are expected from this cross.

 ✓ *Quick check 1*

 ## Quick check question

1 A red-flowered and a pink-flowered snapdragon are crossed. Draw a genetic diagram to show the genotypes and phenotypes of the parents and offspring.

Genetics: dominance and sex linkage

The inheritance of albinism in mice is also controlled by two alleles. The allele, **A**, for coloured fur and coloured eyes is **dominant** to the recessive allele, **a**, for white fur and pink eyes. Three genotypes but only two phenotypes are possible:

genotype	phenotype
AA	coloured fur and eyes
Aa	coloured fur and eyes
aa	white fur and pink eyes (albino)

Consider a cross between a pure-breeding (homozygous) mouse with coloured fur and an albino mouse:

parental phenotypes	coloured fur and eyes	white fur and pink eyes
parental genotypes	**AA**	**aa**
genotype of gametes	all (A)	all (a)
genotype of offspring (F₁)	all **Aa**	
phenotype of offspring (F₁)	all with coloured fur and eyes	

> The symbols for the alleles follow the convention for dominant and recessive alleles of using an upper case letter for the dominant allele and a lower case letter for the recessive.

Now interbreed the heterozygous offspring (F₁ × F₁):

genotypes of F₁	**Aa**	**Aa**
genotype of gametes	(A) or (a)	(A) or (a)
	in equal proportions	in equal proportions

> It is not a good idea to use C/c or W/w for the different alleles in this cross because the upper and lower case letters may easily be confused. P/p and S/s are other letters to avoid. Do not use X or Y because these letters are used for the sex chromosomes.

genotypes and phenotypes of F₂

gametes	**A**	**a**
A	**AA** coloured fur and eyes	**Aa** coloured fur and eyes
a	**Aa** coloured fur and eyes	**aa** albino

This results in one of the classic Mendelian ratios: 3 mice with coloured fur and eyes : 1 albino.

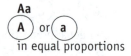

✓ *Quick check 1*

The inheritance of sex and sex linkage

In most animal species, sex is determined by the inheritance of the **sex chromosomes: X** and **Y**. Unlike the other pairs of chromosomes (**autosomes**), X and Y chromosomes are not the same size. They are homologous for only part of their length and this homologous region allows them to pair during meiosis. In mammals, the Y chromosome is much shorter than the X. The X chromosome carries a large number of genes while the Y has very few. A person with a Y chromosome is male. A female (XX) has two alleles of all the genes carried on the X chromosome. A recessive allele will affect the phenotype only if the individual is homozygous for that allele. Much of the X chromosome in a male (XY) is unmatched, because the Y chromosome is so small. A recessive allele of a gene on the X chromosome can affect the phenotype because it is the **only** allele of that gene present.

One of the genes, **H/h**, on the X chromosome codes for the production of a blood clotting agent, **factor VIII**. The dominant allele, **H**, codes for the production of normal factor VIII, whilst the recessive allele, **h**, results in lack of the factor and results in the disease **haemophilia A**.

A haemophiliac man (X^hY) and a normal woman (X^HX^H) have children:

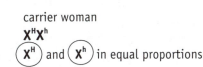

parental phenotypes	haemophiliac man	normal woman
parental genotypes	X^hY	X^HX^H
genotypes of gametes	X^h and Y in equal proportions	all X^H

> The alleles of sex-linked genes are shown as superscript to an X (the chromosome) using upper case (dominant) or lower case (recessive) letters.

genotypes and phenotypes of offspring

		gametes from woman X^H
gametes from man	X^h	X^HX^h carrier female
	Y	X^HY normal male

Now suppose that a woman carrier of haemophilia and a normal man have children:

parental phenotypes	normal man	carrier woman
parental genotypes	X^HY	X^HX^h
genotypes of gametes	X^H and Y in equal proportions	X^H and X^h in equal proportions

genotypes and phenotypes of offspring

		gametes from woman	
		X^H	X^h
gametes from man	X^H	X^HX^H normal female	X^HX^h carrier female
	Y	X^HY normal male	X^hY haemophiliac male

Sons inheriting X^h are haemophiliacs; daughters inheriting X^h are carriers.

✓ *Quick check 2*

❓ Quick check questions

1 The size of the external ear in mice is controlled by a gene with two alleles. Allele **B** gives normal sized ears and allele **b** short ears. A pure-bred mouse with normal sized ears is crossed with a short-eared mouse and the offspring are interbred. Draw a genetic diagram to show the genotypes and phenotypes of the parents, F_1 and F_2 generations.

2 Rarely, a woman carrier of haemophilia and a haemophiliac man have children. Draw a genetic diagram to show the genotypes and phenotypes of the offspring that could be produced.

Genetics: multiple alleles and dihybrid crosses

The crosses shown so far have involved only two alleles of a gene. Most genes have more than two alleles, ranging from a few to very many. Several of the genes controlling the cell surface glycoproteins have over 20 alleles and one has more than 40. Increasing the number of alleles of a gene increases the number of possible phenotypes. But, however many alleles exist of a gene, remember that a diploid organism can have only **two** of them, one on each of the homologous chromosome pair concerned. Two examples of **multiple alleles** are given below.

The gene controlling the ABO blood groups in humans has three alleles producing four blood groups: A, B, AB and O. Alleles I^A and I^B are codominant and I^o is recessive to both. The possible genotypes and phenotypes are shown in the table below.

genotype	blood group (phenotype)
$I^A I^A$	A
$I^A I^o$	A
$I^B I^B$	B
$I^B I^o$	B
$I^A I^B$	AB
$I^o I^o$	O

> Do *not* write the genotypes as AA, AB, AO etc.

> Note that I^A and I^B have upper case letters for the codominant alleles; The recessive I^o has a lower case 'o'.

✓ *Quick check 1*

Fur colour in rabbits is controlled by a gene, **C**, with four alleles showing a hierarchy of dominance.

- The allele, C^A, for agouti (normal) fur is dominant to all other alleles.

- The allele, C^g, for chinchilla (grey) fur is recessive to agouti but dominant to the other two alleles.

- The allele, C^h, for himalayan (white with dark extremities, see page 47) fur is recessive to chinchilla but dominant to the remaining allele.

- The allele, C^a, for white (albino) fur is recessive to all other alleles.

✓ *Quick check 2, 3*

Dihybrid crosses

A **dihybrid cross** looks at the inheritance of **two** genes. Look back at the inheritance of the alleles **A/a** for fur colour in mice (see page 40) and now consider the inheritance of both that gene and one, **B/b**, controlling size of the external ear. The allele **B** codes for normal sized ears and the allele **b** for short ears.

Consider a cross between a pure-bred (homozygous) mouse with coloured fur and normal sized ears and an albino mouse with short ears:

parental phenotypes	coloured fur and normal sized ears	albino and short ears
parental genotypes	**AABB**	**aabb**
genotypes of gametes	all (**AB**)	all (**ab**)
genotypes of offspring (F₁)	**AaBb**	
phenotypes of offspring	coloured fur and normal sized ears	

> The haploid gametes must have one allele of *each* of the two genes.

> Write the genotype of the F₁ AaBb, *not* ABab or aAbB

Now interbreed the offspring (F₁ × F₁):

genotypes of F₁	**AaBb**	**AaBb**
genotypes of gametes	(**AB**) (**Ab**) (**aB**) (**ab**)	(**AB**) (**Ab**) (**aB**) (**ab**)
	in equal proportions	in equal proportions

gametes	AB	Ab	aB	ab
AB	**AABB** coloured fur normal sized ears	**AABb** coloured fur normal sized ears	**AaBB** coloured fur normal sized ears	**AaBb** coloured fur normal sized ears
Ab	**AABb** coloured fur normal sized ears	**AAbb** coloured fur short ears	**AaBb** coloured fur normal sized ears	**Aabb** coloured fur short ears
aB	**AaBB** coloured fur normal sized ears	**AaBb** coloured fur normal sized ears	**aaBB** albino fur normal sized ears	**aaBb** albino fur normal sized ears
ab	**AaBb** coloured fur normal sized ears	**Aabb** coloured fur short ears	**aaBb** albino fur normal sized ears	**aabb** albino fur short ears

phenotypes of offspring: this gives another classic Mendelian ratio of 9:3:3:1.

- 9/16 showing both dominant alleles – coloured fur and normal sized ears;
- 3/16 showing one dominant allele – coloured fur but short ears;
- 3/16 showing the other dominant allele – normal sized ears but albino fur;
- 1/16 showing both recessive alleles – albino fur and short ears.

✓ *Quick check 4*

? Quick check questions

1 State the blood groups that may be expected in the children of a mother with blood group AB and a father with blood group O.

2 List the ten possible genotypes for rabbit fur colour involving the gene, **C**, and give their phenotypes.

3 Two agouti rabbits produce a litter of offspring of which five are agouti and one himalayan. What are the possible genotypes of the agouti parents?

4 Stem colour in tomatoes is controlled by two alleles of a gene. The dominant allele, **A**, gives purple stems and the recessive allele, **a**, gives green stems. Another gene with two alleles controls leaf shape. The dominant allele, **B**, gives leaves with jagged edges and the recessive allele, **b**, gives smooth-edged leaves. A pure-breeding purple-stemmed plant with jagged-edged leaves was crossed with a pure-breeding green-stemmed plant with smooth-edged leaves. The resulting F₁ generation was interbred to give an F₂ generation. Draw a genetic diagram to show the genotypes and phenotypes of the parents and the offspring of both generations.

Genetics: test crosses and χ^2 (chi-squared) test

A **test cross** checks whether an individual is or is not heterozygous at the locus concerned: it is a test of **heterozygosity**. Because of dominance, mice with genotypes **AA** and **Aa** have coloured fur and eyes (see pages 40 and 43). The genotypes cannot be distinguished by examining the phenotypes. Crossing such mice with albinos (**aa**) provides a test cross.

✓ *Quick check 1*

parental phenotypes	coloured fur × albino	coloured fur × albino
parental genotypes	**AA** **aa**	**Aa** **aa**
genotypes of gametes	all (A) all (a)	(A) or (a) in equal all (a) proportions
genotypes and phenotypes of offspring	all **Aa** with coloured fur	1 **Aa** with coloured fur : 1 **aa** albino

Albino offspring can only appear (except by mutation) when the coloured mouse is heterozygous.

Similarly, homozygous and heterozygous mice with coloured fur and normal sized ears, **AABB** and **AaBb**, may be distinguished by crossing with the doubly homozygous recessive, **aabb**:

> ▶ Mutation is so rare that it can almost always be excluded as an explanation.

parental genotypes	**AABB** × **aabb**	**AaBb** × **aabb**
phenotypes of offspring	all with coloured fur and normal sized ears	1:1:1:1
		1 coloured fur and normal sized ears:
		1 coloured fur and small ears:
		1 albino fur and normal sized ears:
		1 albino fur and small ears

✓ *Quick check 2, 3*

χ^2 (chi-squared) test

The χ^2 test is a way of estimating the probability that differences between observed and expected results are due to chance. The formula for χ^2 is:

$$\chi^2 = \text{the sum of } \frac{(\text{observed number} - \text{expected number})^2}{\text{expected number}} = \Sigma \frac{(O - E)^2}{E}$$

The results of a cross in which a 1 : 1 ratio of phenotypes was expected gave 43 offspring with one phenotype and 57 with the other. Is such divergence from expectation due to chance?

The calculation of χ^2 is shown below:

	one phenotype	the other phenotype
O	43	57
E	50	50
O – E	–7	7
$(O - E)^2$	49	49
$(O - E)^2 / E$	0.98	0.98
$\Sigma (O - E)^2/E = \chi^2$	1.96	

Now look at the table below, which shows part of a table of χ^2 values and probabilities for different **degrees of freedom**. The number of degrees of freedom, η, takes account of the number of comparisons made. It is calculated as (number of classes of data – 1). Here, the calculation involves two classes of data, so $\eta = (2 - 1) = 1$.

degrees of freedom (η)	probability, p				
	0.10	0.05	0.02	0.01	0.001
1	2.71	3.84	5.41	6.64	10.83
2	4.61	5.99	7.82	9.21	13.82
3	6.25	7.82	9.84	11.35	16.27
4	7.78	9.49	11.67	13.28	18.47

The probability of $\chi^2 = 1.96$ is > 0.1. In this case the observed results of 43 : 57 are due to chance.

When the difference between the observed and expected results is so large that it would be expected to occur by chance in *fewer* than 1 in 20 experiments (a probability of 0.05), the result is said to **differ significantly** from expectation. The prediction on which the expected results were based is not upheld.

✓ *Quick check 4, 5*

You will always be given the formula for χ^2, or even the value. You need to know what to do with it and what it tells you about the results.

? Quick check questions

1 Explain the purpose of a test cross.

2 Draw genetic diagrams to show the gametes and the genotypes and phenotypes of the offspring of the dihybrid test crosses given above (AABB × aabb and AaBb × aabb).

3 State what cross has been performed to result in each of the following phenotypic ratios: (i)1 : 1; (ii) 1 : 1 : 1 : 1.

4 The results of a cross in which a 3:1 ratio of phenotypes was expected gave 160 offspring with one phenotype and 40 with the other. Use the χ^2 test to find whether the divergence from expectation is due to chance.

5 The results of a dihybrid test cross, in which a 1 : 1 : 1 : 1 ratio of phenotypes was expected, gave the following numbers of four phenotypes: 70, 65, 35, 30. Use the χ^2 test to find whether the divergence from expectation is due to chance or whether the expected ratio was based on an incorrect assumption.

Mutation and the phenotype

A **mutation** is an unpredictable change in the genotype of a cell that results from damage to, or imperfect replication of, the DNA or faulty operation of mitosis or meiosis. Such a change may be a **gene mutation** or a **chromosome mutation**.

- Mutation is a very rare event. Most genes have less than one chance in a million of being incorrectly replicated.

- Mutation takes place in a single cell but cannot usually be detected until the mutant cell has replicated and a distinctive phenotype has been produced.

- Some chemicals, **mutagens**, increase the chance of mutation because they react with DNA or interfere with replication.

- Some types of radiation, e.g. X-rays, cause mutation by damaging DNA.

A **gene mutation** affects the nucleotide sequence of a section of DNA so that the triplet code for the primary structure (the amino acid sequence) of the polypeptide coded by the gene is changed.

> Look up the genetic control of protein structure from your AS course.

- One or more nucleotides may be **deleted** from, or **added** to, the sequence. This alters the reading frame of the triplet code, altering all subsequent triplets.

- One nucleotide may be **substituted** for another, e.g. guanine instead of thymine, producing a different triplet code which may code for a different amino acid.

- A triplet sequence may be repeated many times (a 'stutter').

> ✓ *Quick check 1, 2*

A **chromosome mutation** alters the **number** of chromosomes in a cell or the **structure** of chromosomes, so altering the sequence of genes in one or more chromosomes.

The genetic disease **sickle cell anaemia** is an example of a **substitution** mutation. All the symptoms of sickle cell disease result from *one* nucleotide substitution in the triplet code for *one* amino acid. Close to the end of the gene coding for the β-globin chain in haemoglobin, a triplet in the normal allele codes for the amino acid glutamic acid. In the mutated allele the coding is for the amino acid valine.

> Look up the structure of haemoglobin from your AS course.

```
                              triplet code
                              C-T-T
normal β-globin chain    .............. Lys-Glu-Glu-Pro-Thr-Leu-His-Val (end of chain)
sickle cell β-globin chain  .............. Lys-Glu-Val-Pro-Thr-Leu-His-Val
                              C-A-T
                              triplet code
```

The presence of valine, which has a hydrophobic R group, allows the sickle cell haemoglobin (HbS) to bind to adjacent HbS molecules. Many HbS molecules can stick together in long tubules in the red cell, distorting the cell into a sickle shape. Sickle cells carry less oxygen, get stuck in capillaries, obstructing blood flow, and are destroyed by phagocytes. Many tissues and organs of the body have a severe lack of oxygen: the person suffers sickle cell anaemia.

> ✓ *Quick check 3*

The genotypes and phenotypes with respect to this gene are shown below. H^N is the allele for normal β-globin and H^S the allele for sickle cell β-globin.

Note that the symbols for the alleles follow the convention for codominant alleles.

genotype	phenotype
$H^N H^N$	normal
$H^N H^S$	sickle cell trait: half normal and half sickle cell haemoglobin. Problems only arise during very strenuous exercise or at high altitude
$H^S H^S$	sickle cell anaemia

The presence of the allele H^S in the genotype affects susceptibility to malaria: the heterozygote has a selective advantage in malarial areas compared with the homozygote $H^N H^N$ (see page 55).

Phenotypic variation

Genotype and environment contribute to **phenotypic variation**. Their interaction can be written as: $V_P = V_G + V_E$

where V_P = phenotypic variation; V_G = the genetic component of variation and V_E = the environmental component of variation.

Environmental effects may allow the full genetic potential to be reached or may in same way restrict it. A pea plant with the dominant allele for tallness may grow only to the same height as a plant homozygous for the recessive allele for shortness because it is in a soil with few nutrients.

Environmental variation (V_E) cannot be inherited: it is the result of differences in the environment experienced during an individual's lifetime. It is important to recognise the extent of environmental variation when selectively breeding organisms. Because environmental variation cannot be inherited it cannot be selected for (see page 54).

The recessive allele producing aerodynamically unsatisfactory **curled wings** in *Drosophila* is expressed only if the pupae are kept at a higher temperature than normal (27°C) for the last day of their pupal life.

In the **himalayan** colouring of rabbits (see page 42) and in Siamese and Burmese cats, the allele for the formation of dark pigment is only expressed at low temperatures. The extremities, e.g. the ears, nose and paws, are cooler than the core temperature of the body, so the allele is able to have its effect in these cooler regions.

✓ *Quick check 4,5*

Quick check questions

1 Explain how the deletion or addition of one nucleotide into the triplet code of DNA affects polypeptide production.

2 Explain why the substitution of one nucleotide for another in a length of DNA may have no effect on the polypeptide produced.

3 Describe the consequences of a person inheriting the genotype $H^S H^S$.

4 Explain why the allele producing himalayan colouring in rabbits affects only the extremities of the animal.

5 Predict the phenotypes of two groups of genetically identical dwarf pea plants when one group is grown in high, and one in low light intensity but in otherwise identical environmental conditions.

Questions 1 and 2 are synoptic. Revise protein synthesis from your AS course.

Regulation of gene activity

The enzyme β-**galactosidase** hydrolyses the disaccharide **lactose** to the monosaccharides glucose and galactose.

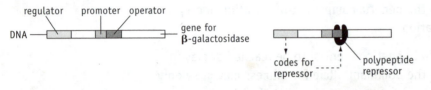

$$\text{lactose + water} \xrightarrow{\text{β-galactosidase}} \text{glucose + galactose}$$

In the bacterium *Escherichia coli* the number of molecules of the enzyme present in the bacterial cell varies according to the concentration of lactose in the medium in which the bacterium is growing. The bacterium has one copy of the gene for β-galactosidase and so must regulate the transcription of that gene in order to alter the concentration of the enzyme in its cytoplasm.

A length of DNA adjacent to the gene for β-galactosidase contains regulator, operator and promoter regions, as shown in the figure. The sequence is known as the *lac* **operon**.

The sequence of events in the *absence* of lactose in the bacterial growth medium is as follows:

- the polypeptide coded by the regulator gene is a repressor;
- the repressor binds to the operator region close to the β-galactosidase gene;
- in the presence of bound repressor at the operator, RNA polymerase cannot bind to the DNA at the promoter region;
- no transcription of the gene for β-galactosidase occurs.

When lactose is *present* in the bacterial growth medium:

- lactose is taken up by the bacterium;
- lactose binds to the repressor molecule, distorting its shape and preventing it from binding to DNA at the operator site;
- transcription is no longer inhibited and β-galactosidase is produced.

> This is similar to the inhibition of an enzyme by distortion of the molecule.

> ✓ *Quick check 1*

In this way, the bacterium produces β-galactosidase only when lactose is present in the surrounding medium. It avoids the waste of energy in producing an enzyme for hydrolysing a sugar it may never encounter. But when the sugar is available it can be hydrolysed.

> ✓ *Quick check 2*

? Quick check questions

1 Explain why preventing RNA polymerase from binding to DNA prevents transcription.

2 State the advantages of a mechanism (such as the *lac* operon) for switching on a gene only in certain circumstances.

> Question 1 is synoptic. Revise mRNA from your AS course.

The Human Genome Project

The Human Genome Project is an international collaboration involving governments, medical charities and pharmaceutical companies to produce a complete sequence of the DNA bases of every human chromosome. The sequence of almost all (94%) of the genome was published in February 2001.

The complete sequence of chromosome 22 (except for 11 small gaps) was published in December 1999, and that of chromosome 21 (except for 3 gaps) in May 2000. These chromosomes are among the smallest of the human chromosomes. Chromosome 22 has over 500 genes and there may be as many as 1000, but chromosome 21 has only 225 identified genes. At least 27 human disorders are associated with genes on chromosome 22.

Implications of sequencing the human genome

The sequence is simply a string of letters (A, T, C and G) denoting the bases in the sequence, but knowing the sequence should:

- speed up identification of genes;
- stimulate research to discover the function of newly discovered genes;
- improve the understanding of how each gene works, individually and with others;
- lead to amino acid sequences and so to more polypeptide structures and functions;
- lead to drugs that could affect these polypeptides;
- allow more gene probes to be produced for genetic testing;
- enable doctors to anticipate and take precautions against disease in an individual;
- allow comparison of the human genome with those of other organisms whose sequence is known, e.g. bacteria, yeast, nematode worm and *Drosophila*;
- allow identification of the small differences between one individual's code and another's, allowing drugs to be tailor-made for each individual genome, avoiding many drug side-effects.

Problems arising from DNA sequencing

The Human Genome Project is making the DNA sequence freely available. Private companies working outside the project regard the sequences as commercial property. A fee is then payable for their use, e.g. in developing gene tests.

In the UK it became possible in October 2000 for insurance companies to ask anyone who has taken a gene test for Huntington's disease, and who is seeking life insurance, to reveal the result of the test. Permission is being sought for the results of tests for nine other diseases, including Alzheimer's, hereditary breast cancer and myotonic dystrophy, to be revealed. This may discourage people at risk of a disease from taking a test that could give early diagnosis and treatment.

Lengths of DNA are measured in thousands of bases (kilobases) or millions of bases (megabases). The human genome contains an estimated 30 000 genes, occupying less than 5% of the 3000 megabases of human DNA. The remaining DNA consists of non-coding regions whose function is largely unknown: the so-called 'junk' DNA.

Human chromosome pairs are numbered in order of size from 1 (large) to 22 (small). The sex chromosomes X (large) and Y (small) are numbered 23.

Genes with no known function are called 'orphan' genes.

Remember that each person's DNA is unique, and regions of DNA differ from one person to another. Single nucleotide polymorphisms (SNPs), which are changes of single base pairs, are relatively common in the human genome.

Exam-style questions

1 Groundsel is a common weed. Groundsel plants vary in the structure of their flowers. Three distinctive phenotypes can be recognised and are shown in the figure.

long ray   short ray   no ray

magnification ×1.5

All flowers on any one plant are always of the same type.

Seedlings of groundsel were collected from a population known to contain all three types. The seedlings were grown to maturity in individual pots under conditions where cross-pollination would be unlikely. The seeds produced by each of the resulting plants were collected and sown separately. When the offspring had grown to maturity, each was examined and its flower type noted. The results are shown in the table below.

phenotype of the parent plant	number of offspring of each parent with the phenotype		
	long ray	short ray	no ray
long ray	103	0	0
long ray	89	0	0
short ray	38	87	41
long ray	23	0	0
no ray	0	0	101
long ray	121	1	0
no ray	0	0	76
short ray	53	111	61
long ray	153	0	0
short ray	23	42	19

With reference to the table:

   a   describe the trends shown by the data;   [3]

   b   explain the results using suitable genetic symbols and diagrams;   [5]

   c   predict the offspring that would result from a cross between a long-rayed plant and one with no rays;   [1]

   d   state **two** hypotheses to explain the single short-rayed offspring produced by parent number 6; assume this plant was correctly identified as short rayed.   [2]

2 Mature, ripe, pea seeds vary in shape. Some commercial varieties have smooth and almost round seeds. Other varieties have more irregular wrinkled seeds. This variation is inherited and the pattern of inheritance shows that one gene is involved. A dominant allele controls the development of round seeds and a recessive allele controls the development of wrinkled seeds.

A plant with wrinkled seeds was cross-pollinated and one of its offspring had round seeds.

◖ This question is synoptic.

a Predict the ratio of offspring that would be produced if this round-seeded plant was cross-pollinated by a plant with wrinkled seeds. [1]

A round seed and a wrinkled seed were crushed in water and a random sample of starch grains from each seed was observed using a microscope. The grains from the round seed were usually oval in shape while most of those from the wrinkled seed were triangular. The grains were measured at their widest point. The data are recorded in the table below.

| maximum diameter of starch grain/μm | from round seed | 32 | 29 | 18 | 23 | 22 | 31 | 28 | 26 | 21 | Mean = |
| | from wrinkled seed | 6 | 7 | 5 | 7 | 7 | 5 | 8 | 5 | 6 | Mean = |

b With reference to the table, calculate the mean diameter for each type of starch grain and insert the means in the spaces provided in the table. [2]

c Suggest how random grains might be selected for measurement using a microscope. [1]

d State how the grains could be identified as starch. [1]

e Explain how it is possible for the size of starch grains in a seed to vary, even though all cells in the seed have the same genotype. [2]

The figure shows the formula of glucose-1-phosphate.

f With reference to the figure, state whether the glucose in this substance is alpha or beta glucose. [1]

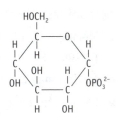

A starch-free solution, rich in soluble proteins, can be extracted from pea seeds. A few drops of this extract will produce starch if added to a solution of glucose-1-phosphate. No starch is produced by the extract if it is added to glucose or to maltose solutions.

g Suggest how it could be demonstrated that pea seed extract contains an enzyme that controls the formation of starch from glucose-1-phosphate. [1]

h Assuming that starch is synthesised from glucose-1-phosphate within developing pea cells, suggest what substance reacts with glucose to produce the glucose-1-phosphate in the cells. [1]

Extracts made from round pea seeds are always more active in the synthesis of starch than extracts made from wrinkled peas.

i Describe **two** experimental variables that would need to be controlled when comparing the enzyme activity of extracts from round and wrinkled peas. [2]

j Explain the relationship between the enzyme activity of pea seeds and the genotype of the seeds. [2]

Classification, Selection and Evolution

It is difficult to produce a definition of the term **species** that satisfies all biologists. Two definitions are used.

- A group of more or less similar organisms capable of interbreeding and producing fertile offspring and that are reproductively isolated from other such groups.

- A group of organisms showing a close similarity in a number of characteristics (morphological, physiological, embryonic, ecological and behavioural).

✓ Quick check 1

The first definition, that of a 'good' species or **biospecies**, excludes:

- organisms that reproduce asexually or by parthenogenesis;

- organisms known only from preserved specimens (**morphospecies**);

- fossils (**palaeospecies**);

- single specimens;

- **ring species**, because individuals from opposite extremes of the range may not interbreed.

❷ Parthenogenesis is the development of an *unfertilised* gamete into a new individual.

✓ Quick check 2

Biospecies do not breed successfully with members of other species. The isolating mechanisms take different forms.

- Members of different species may have different appearances or behaviour patterns so that courtship does not begin or is not completed.

- Mating may not be physically possible.

- Fertilisation may fail or the zygote fail to develop.

- The species may reproduce at different times or in different places.

- Interbreeding may occur but the hybrids are infertile, e.g. mules.

Species are named according to the **binomial** system introduced by Carl Linnaeus in about 1758. In this system a species is given a two-part latinised name: the name of the genus and the name of the species, e.g. *Homo sapiens*.

❷ Note the upper case letter for the genus, the lower case letter for the species and the italics.

Taxonomic groups

Species are classified into groups that show many similarities. These groups of species are in turn classified into fewer, larger, groups that have fewer but more fundamental similarities. All classifications attempt to identify important similarities and differences rather than those due to chance. Similarities and differences that result from evolutionary relationships are the most fundamental. The history of an organism's evolutionary relationships is called its **phylogeny**.

❷ Classificatory groups are called **taxa** (singular: taxon).

Features being compared when establishing evolutionary relationships can be homologous or analagous.

- **Homologous** structures are thought to have an evolutionary origin in the same ancestral structure, e.g. the pentadactyl limbs of different classes of chordate all have a humerus bone. Protein molecules can also show homology of the amino acid sequence, e.g. the cytochromes of the electron transport chain.

● **Analagous** structures have evolved independently and develop in a different way to carry out the same function, e.g. the wings of insects and birds, and the eyes of squids and fish.

A **taxon** is a group of organisms classified together. There are seven levels of taxon:

taxon	number of similarities	size of group	degree of relatedness	example
kingdom	smallest number of similarities	largest group	distantly share a common ancestor	Animalia
phylum				Chordata
class				Mammalia
order				Primates
family				Hominidae
genus				*Homo*
			share a common ancestor increasingly recently	
species	largest number of similarities	smallest group		*sapiens*

A major problem in taxonomy is how many kingdoms, the largest taxa, to use to group living organisms. Sometimes only two (or three), Prokaryotae and Eukaryotae (Akaryotae for the viruses), are used, but it is recommended that you use **five** kingdoms. Members of the kingdoms differ in the organisation of their cells.

✓ *Quick check 3, 4*

❶ Remind yourself of the characteristics of prokaryotic and eukaryotic cells from your AS course.

kingdom	features	examples
Prokaryotae	Cells lack nuclei organised within nuclear envelopes, envelope-bound organelles and microtubules.	Bacteria, including the photosynthetic cyanobacteria.
Protoctista	Eukaryotes; unicells or assemblages of similar cells. Some are autotrophs, some heterotrophs.	All eukaryotic organisms that are not fungi, plants or animals. All nucleated algae, all protozoa and slime moulds.
Fungi	Eukaryotes with cell walls not of cellulose but usually of chitin. They are usually organised into multinucleate hyphae and are non-photosynthetic (heterotrophs) with absorptive methods of feeding.	Mushrooms, moulds such as *Penicillium* and yeasts.
Plantae	Eukaryotic organisms that are multicellular photosynthetic autotrophs and have cellulose cell walls. Plants are non-motile.	Liverworts, mosses, ferns, conifers and flowering plants.
Animalia	Eukaryotic organisms that are multicellular and non-photosynthetic (heterotrophs) with nervous co-ordination.	All groups from jellyfish to mammals.

The viruses **(Akaryotae)** cannot be fitted into the five kingdom classification.

✓ *Quick check 5*

? *Quick check questions*

1 State two definitions of the term species.

2 Explain why not all species are biospecies.

3 List the hierarchy of taxa used in biological classification.

4 Explain the link between classification and evolutionary relationship.

5 Explain why the viruses cannot be fitted into the five kingdom classification.

❶ You will learn more about viruses, prokaryotes, protoctists and fungi if you study Microbiology and Biotechnology as your optional module.

Changing allele frequencies in populations

The processes that act on a natural population of organisms to alter allele frequencies fall into two categories:

- random processes, e.g. **genetic drift**;
- non-random processes, e.g. **selection**.

Genetic drift

Genetic drift is a change in allele frequency that occurs **by chance**, because only some of the organisms in each generation reproduce. Genetic drift is particularly noticeable when a *small* number of individuals are separated from the rest of a large population. They form a small sample of the original population and so are unlikely to be representative of the large population's gene pool. Genetic drift in the small population will alter the allele frequencies still further. This process, occurring in a recently isolated small population, is called the **founder principle**.

Selection

Genetic variation (and so phenotypic variation) in a population of organisms means that some organisms will have a better chance of survival than others. They are better adapted to the particular conditions and are more likely to breed successfully and contribute more offspring to the next generation than their less fit competitors. This **natural selection**:

- increases the chance of advantageous adaptive alleles being passed to the next generation;
- decreases the chance of non-adaptive alleles being passed to the next generation.

An individual's **fitness** is a measure of its ability to transmit its alleles to the next generation as a result of natural selection.

In unchanging conditions, **stabilising selection** maintains existing adaptations and so maintains existing allele frequencies.

In changing conditions **directional selection** alters allele frequencies.

A new allele (a **mutation**) appearing in the population may be disadvantageous in existing conditions and be removed by stabilising selection. But should conditions change, the new allele might be advantageous and selected for. Selection then becomes an evolutionary force.

Stabilising selection is seen best where the heterozygote for a particular gene has a selective advantage and the two homozygotes have selective disadvantages. This maintains all genotypes in the population because it tends to be the heterozygotes that breed.

✓ **Quick check 1**

❶ A population's gene pool is the sum total of its alleles.

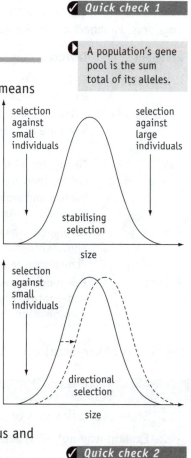

✓ **Quick check 2**

❶ Do not confuse evolutionary fitness and physiological fitness, as used in sports physiology.

- In malarial areas, humans heterozygous for **sickle cell haemoglobin** (see page 47) are resistant to malaria and do not show the symptoms of sickle cell anaemia. Individuals homozygous for the normal allele are susceptible to malaria and those homozygous for sickle cell haemoglobin suffer from sickle cell disease. Until recently, sufferers of sickle cell anaemia usually died before reaching reproductive age. The heterozygotes, who do not suffer from sickle cell anaemia and are much less likely to suffer badly from malaria, contribute more children to the next generation. The expected ratio of genotypes when two heterozygotes have children is 1 homozygote dominant : 2 heterozygotes : 1 homozygote recessive. Exactly the same selective pressures that applied to the previous generation now apply, providing that this new generation is still in a malarial area.

- The rat poison **warfarin** interferes with vitamin K metabolism and so prevents blood clotting. The rats die from internal bleeding. Resistance or susceptibility to warfarin is controlled by a single gene with two codominant alleles: one giving resistance and one susceptibility. In areas where warfarin is used to kill rats, those that are homozygous for the allele for warfarin susceptibility die. Both the heterozygotes and those homozygous for the allele for resistance are resistant to warfarin, but the homozygotes need large quantities of vitamin K in their diet and die if they do not have it. Again, the heterozygotes tend to contribute more offspring to the next generation, maintaining both alleles in the gene pool.

Directional selection is seen best in changing conditions, e.g. pollution favouring **industrial melanism** in *Biston betularia*, the peppered moth.

Until the mid-19th century the peppered moth was found as its *typical* variety with pale speckled wings, camouflaging the adult insect from bird predators when it rested by day on lichen-covered tree trunks and branches. With increasing air pollution, a black **(melanic)** form, called *carbonaria*, produced by a dominant allele was better camouflaged on soot-covered lichen-free trees. *Typical* moths were easily detected and eaten, *carbonaria* increased in the population and the frequency of the dominant allele rose in polluted areas. The population of *typical* peppered moths before the industrial revolution was homozygous recessive. The dominant *carbonaria* allele arose by chance mutation and, because it had a selective advantage in polluted areas, its frequency increased in those areas. The dominant allele also affects the viability of the caterpillars: caterpillars that will become *carbonaria* moths are more tolerant of eating polluted vegetation. So in a polluted area more caterpillars carrying the dominant allele will survive to pupate (although the caterpillars are not melanic). The effect of the *carbonaria* allele has changed with time; *carbonaria* moths are darker than they once were, which shows that other genes that enhance the effect of the *carbonaria* allele are being selected for.

> Remind yourself about malaria from your AS course.

> ✓ Quick check 3

> Remember the convention for codominant alleles. Rw represents the gene. RwR and RwS are the alleles for resistance and susceptibility, respectively.

> ✓ Quick check 4

> A mutant dominant allele is seen in the phenotype and may be immediately subjected to selection. A recessive allele must be homozygous before it can be selected for or against.

> ✓ Quick check 5

Quick check questions

1 Explain what is meant by the term genetic drift.
2 Distinguish between stabilising selection and directional selection.
3 Draw a genetic diagram to show the expected ratio of genotypes among the children of two individuals heterozygous for sickle cell haemoglobin (H^NH^S).
4 Explain how mating between rats heterozygous for resistance to warfarin (Rw^RRw^S) maintains both alleles in a population on which warfarin is used.
5 Explain why the *carbonaria* allele is found at high frequency in populations of peppered moths in polluted areas, but not in rural areas.

Speciation

Speciation is the evolution of two or more groups that do not interbreed and are therefore new species. There are two ways in which one group of interbreeding organisms can produce another group that cannot interbreed with the first.

- **Allopatric speciation**: this requires geographical isolation and time and involves a combination of genetic drift and natural selection.

- **Sympatric speciation**: this occurs when the organisms are in the same place at the same time. Speciation can occur in plants in a single generation as a result of **polyploidy**: the different chromosome numbers of parents and offspring provide reproductive isolation.

> n = haploid;
> $2n$ = diploid;
> $4n$, $8n$ etc. = polyploid.

Allopatric speciation

Isolated islands provide the ideal scenario for allopatric speciation.

- Imagine a population of organisms, **species A**, with a niche on a particular island, **A**.

- By chance, a few breeding individuals are dispersed to a distant, isolated island, **B**, where the conditions are slightly different.

- Some of the individuals are sufficiently well-adapted to these different conditions to survive and breed.

- Because the population is small, genetic drift occurs, altering allele frequencies.

- Natural selection occurs, selecting for different traits, and hence different alleles on each island.

- With each generation, the two populations differ increasingly.

At this stage, if the two populations were recombined on island **A**, it is unlikely that they would have, by chance, any reproductive barriers. The two populations would hybridise and the resulting population might well look different from both species **A** and **B**. The population would remain one species.

However, suppose that rarely a few individuals from **A** are dispersed to **B** and *vice versa*. Offspring of matings between individuals from **A** and **B** would be at a disadvantage on both islands in comparison with matings between members of each island's own population. Some form of recognition between members of the same population is likely to be **selected** for. The two groups behave as two 'good' species: **species B** has evolved on island **B**.

If the two islands are so widely separated that no such chance migrations between the two populations occur, the reproductive barriers will **not** be selected for and the two groups may not be 'good' species.

✓ *Quick check 1,2*

This process may be seen in its **intermediate** stages in many island populations isolated from the same species on the nearest mainland. For example, the wren *Troglodytes troglodytes* differs in size and colour on the mainland of Scotland and on the various Scottish islands. As yet, no reproductive isolation exists and so the species remains just one species.

✓ *Quick check 3*

The process described above is the likely explanation for the existence of Darwin's Finches: the 13 species of finches of the Galapagos Islands. All the species have dull brown or black plumage and have similar eggs, nests and courtship behaviour: so similar that it is generally agreed that they share a common ancestor. The main differences between the species are in the shape and size of the beak.

- The seed- and cactus-eating species of **groundfinch** have beaks of the same shape but different sizes, adapted for feeding on different sized seeds or plants so that the species are not in competition with one another.

- **Treefinches** include those with beaks adapted for eating insects and the woodpecker finch which uses a cactus spine to probe for and catch insects in wood.

- The **warbler** finches are adapted for insect-eating.

Finch bill shape and size is an example of a structural adaptation to its environment.

It is probable that a few ancestral finches were blown from the South American mainland to one of the volcanic islands. They survived and became adapted to the conditions by natural selection. A few birds were then carried to another island where they survived and gave rise to a differently adapted population. Provided there were occasional migrations between the two islands, reproductive isolation could be selected for, producing two species. These could then both live on the same island, without interbreeding, provided the island had suitable habitats. Repetition of this process gave rise to the other species. Such **adaptive radiation**, where many species evolve from a single species, was possible because few other species were present on the isolated volcanic islands.

Measurements have been taken of the beak sizes of six species of seed-eating groundfinch on one of the islands. When there is plenty of rain and seeds are abundant, all six species, no matter what their beak size, eat softer, smaller seeds. During a drought in 1977 most plants did not produce seeds and by the end of the year only a few large, hard seeds remained as food. The species with the largest beak size were able to eat these seeds easily, but the seeds were at the limit of what one of the species of medium groundfinch could tackle. Many birds died, but those with larger beaks than others survived, leaving a population of medium groundfinches after the drought with a larger average beak size than before.

Sympatric speciation

Sympatric speciation requires *instant* reproductive isolation. For example, a grass, *Spartina maritima*, grew in salt marshes in southern England. A different species, *S. alterniflora*, was introduced accidentally from America. The two species hybridised to give a *sterile* diploid (2n) hybrid that spread asexually by means of rhizomes. At some point in the 1890s, polyploidy resulted in the chromosome number of the hybrid doubling giving a *fertile* **tetraploid** (4n). This plant was a new species: *S. anglica*. It grows so vigorously that it is now regarded as an invasive weed.

✓ *Quick check 4*

? Quick check questions

1 Explain why isolation is important in speciation.

2 Explain why it is necessary for there to be reproductive barriers between 'good' species.

3 Why must the wrens of the different Scottish islands be considered to be the same species as the mainland wren?

4 Explain how polyploidy provides instant reproductive isolation.

Artificial selection

Artificial selection occurs when a breeder chooses individuals with desirable phenotypes for breeding and/or prevents those with less desirable phenotypes from breeding, thus changing allele frequencies in the population.

Selective breeding

Domesticated animals and crop plants have been subjected to artificial selection for a very long time through **selective breeding**. Individuals showing one or more of the desired traits have been chosen for breeding and, in turn, the most suitable individuals among the offspring have been selected. Over many generations the alleles giving the desired traits have increased in frequency, whilst other alleles have decreased in frequency or been lost.

✓ Quick check 1

Artificial selection may be compared with natural selection:

artificial selection	natural selection
selective agent is the breeder	selective agent is the total environment of the organism
traits of use to the breeder are selected, including those which may not be to the organism's advantage	adaptations to the prevailing conditions are selected
a single trait may be selected	many different traits contributing to fitness are selected

✓ Quick check 2

Aims of artificial selection

The aims of artificial selection include:
- increasing the yield and quality of crop plants;
- producing pest- and pathogen-resistant varieties of crop plants;
- producing varieties of crop plants that are tolerant of changes in temperature or water availability;
- increasing the meat production, milk yield and egg yield of livestock;
- reducing the fat content of meat;
- improving the performance of animals used in competitive sports, e.g. race horses;
- producing new colours of garden flowers;
- producing particular characteristics in breeds of domestic animals, e.g. long hair in cats.

Successful artificial selection depends on choosing parents whose desirable traits are produced largely by their genotype and not by the environment (see page 47). The **heritability** of a trait is a measure of that part of the desirable characteristic of the parent that is, on average, passed to the offspring. Heritability values can vary in theory from 0 (no genetic contribution) to 1.0 (no environmental effect). Hence a trait with low heritability (<0.02) means that selective breeding for that trait will have little effect.

Many of the desirable traits of crop plants and livestock are controlled not by single genes but by many genes: **polygenes**. There is often a large environmental component to the variation seen, which is why selective breeding for such characters is unpredictable. There is no guarantee that parents with desirable characters will produce the wanted offspring.

An example of artificial selection: rice

The main centre for the selective breeding of rice is the International Rice Research Institute (IRRI) in the Philippines. IRRI holds the rice gene bank. By the 1960s semi-dwarf varieties of rice had been bred, giving yields of approximately 10 tonnes per hectare. Much selective breeding since then has been to develop varieties with resistance to pests and diseases and that are more tolerant of poor soils, drought or flooding (rice is not adapted to being totally submerged). A variety released in the 1980s, for example, had resistance to seven important insect pests as a result of a complex breeding programme involving varieties from India, China, Taiwan and the Philippines. Such varieties helped farmers to achieve maximum yields but did not increase that maximum. After a five-year breeding programme, a new variety with higher yield was developed in 1994, capable of yielding 13 tonnes per hectare in optimum conditions. Unlike previous varieties, it produces seed heads on every shoot and has almost double the number of seeds per head (more than 200 grains per head). It is also very compact, so that it can be planted more densely. Since then selective breeding has continued to introduce resistance to pests and diseases to this variety without loss of yield.

It is important that many traditional varieties and wild populations of plants such as rice are conserved to allow future crossing, followed by selection.

> ▶ A gene bank is a store of a species' genetic diversity.

> ✓ Quick check 3

> ✓ Quick check 4

> ▶ You will learn more about heritability and selective breeding if you study Applications of Genetics as your optional module.

? Quick check questions

1 Explain the purpose of artificial selection.

2 Distinguish between natural selection and artificial selection.

3 List the traits that have been artificially selected for in the selective breeding of rice.

4 Many traits that are seen as desirable in crop plants are controlled by polygenes. Suggest what problems this presents in a selective breeding programme.

Exam-style questions

1 Humans (*Homo sapiens*) and chimpanzees (*Pan troglodytes*) have completely different scientific names. Dogs (*Canis familiaris*) and wolves (*Canis lupus*) have one scientific name in common.

   a Explain this difference in classification.   [2]

   Penicillium chrysogenum and *Penicillium notatum* are fungi that produce the antibiotic penicillin. Neither fungus has been observed to reproduce sexually.

   b Explain why this was a problem for taxonomists when these fungi were classified.   [2]

   c Suggest **three** reasons why biologists have spent time and effort on the classification of organisms.   [3]

2 The Orkneys are a group of islands lying to the north of the British Isles, between Scotland and Norway. There are populations of voles on several of these islands. A vole is a small mouse-like rodent.

   Similar voles, classified as belonging to the same species, are found on the Scottish mainland. The Orkney voles are much larger than those found on the mainland, they also have a different coat colour.

   Electrophoresis, of both blood proteins and DNA fragments, from Orkney and mainland voles reveals more fundamental differences between the populations.

   a Explain how both isolation and natural selection have produced these differences.   [7 +1 for quality]

   b Discuss the idea that voles from Orkney and the Scottish mainland belong to different species.   [5]

3 The human Y chromosome is much smaller than the X and has few genes, only one of which has a known function. The rest of the DNA of the Y chromosome does not code for protein and its function, if any, remains unknown.

   One part of the Y has a nucleotide sequence that is homologous with part of the X, allowing the X and Y chromosomes to pair.

   a Describe the circumstances under which pairing happens.   [2]

   Most of the Y chromosome has a nucleotide sequence that is quite different from that of the X, or to that of any other chromosome. This part of the Y chromosome cannot recombine with any other DNA sequence and can only change as a result of mutation. This means that the nucleotide sequence of the Y is much less variable than that of other chromosomes. Large numbers of men have identical Y chromosomes and they must all have descended from a common male ancestor.

   b Discuss the hypothesis that men with the same Y chromosome are likely to be genetically similar in other ways.   [3]

> Questions 2 and 3 use quite long passages that you must understand as fully as possible before you try to answer. The parts of the question will require you to use information or hints provided by the passage together with what you already know. **Always read such a passage through twice before answering any part of the question.**

> Discuss indicates that there are points for and against.

If two men have Y chromosomes that show a single difference in nucleotide sequence, this must have come about as a result of a single mutation in one of the two lines of descent from their common ancestor.

By comparing the DNA of Y chromosomes from many men, from different ethnic origins, it is possible to estimate the total number of mutations that have taken place in the Y chromosome since humans evolved.

Women have no chromosomal DNA that remains free from recombination.

Mitochondrial DNA, however, like that of Y chromosomes, does not recombine. Because men and women inherit all their mitochondria from their mother's egg cells, an analysis of the nucleotide sequence of mitochondrial DNA and a comparison of the mitochondrial DNA from many individuals, allows evolutionary relationships to be worked out in the same way as in the case of the Y chromosome. If there is only one difference in mitochondrial nucleotide sequence then two people have a recent common female ancestor.

Comparison of the mitochondrial DNA of humans (*Homo sapiens*) and chimpanzees (*Pan troglodytes*) shows numerous differences in nucleotide sequence. The similarities, however, are so great that the two species must have shared a common female ancestor much more recently than biologists once believed, when they could only compare anatomy, behaviour and other phenotype differences.

There is much more variation in the nucleotide sequence of mitochondrial DNA in chimpanzees than there is in the sequence of human mitochondrial DNA.

c   Suggest an explanation for the difference given in the last sentence of the passage above.   [2]

d   Explain why recombination is important for evolution.   [3]

e   Explain why the DNA of Y chromosomes is particularly suitable for assessing genetic relationships.   [2]

f   State the meaning of the term *homologous*.   [2]

4   Cereals were domesticated from several wild species of grass. When cereals are compared with their ancestral species, it can be seen that the parts of the ear that contain the mature seeds are strongly attached to the stem in cereals, but are brittle (easily shaken from the stem) in the wild species. This difference is seen in barley, wheat, rice, millet and all cereal species.

Suggest and explain why wild species have brittle ears while domesticated varieties have strongly attached ears.   [5]

Control, Coordination and Homeostasis

Homeostasis is the maintenance of a relatively stable **internal environment** of an organism. This provides relatively constant conditions for the individual cells, even when the external environment fluctuates or the organism changes its behaviour, e.g. by moving faster or feeding.

✓ *Quick check 1*

Homeostasis requires:

- a **receptor** (sensor or detector) that receives information (input);
- a **control mechanism** that responds to information and stimulates an effector;
- an **effector** to perform appropriate action.

input → **receptor** → **control mechanism** → **effector** → output

← negative feedback ←

Negative feedback occurs when a change in a system sets in motion a sequence of events that counteracts the change and restores the original state. For negative feedback to occur there must be an optimal norm **reference point** or **set point**. For example, a rise in mammalian blood temperature sets in motion mechanisms to reduce blood temperature, but when the temperature falls below the norm a new sequence of events raises the temperature. Negative feedback is also seen in mammals in the homeostasis of:

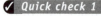

▶ In *positive* feedback a change in a system results in further change away from the original state.

- blood glucose concentration;
- ion concentrations of blood;
- water content of body fluids;
- carbon dioxide concentration of blood.

✓ *Quick check 2*

Excretion

Excretion is the process by which an organism gets rid of waste metabolic products, e.g. nitrogenous waste from amino acid metabolism and carbon dioxide from respiration. These waste products are harmful or **toxic** and would damage cells if allowed to accumulate.

✓ *Quick check 3*

Mammals cannot store proteins or amino acids, so any excess is converted into fats or carbohydrates for storage or use in respiration. The amino group ($-NH_2$) of each amino acid is removed in the process of **deamination** in the liver, forming the very soluble but very toxic compound **ammonia**. This is combined with carbon dioxide from respiration, in the **ornithine cycle** in the liver, to give the much less toxic but adequately soluble compound **urea**. This is removed from the blood by the kidneys.

Carbon dioxide from respiration combines with water to give carbonic acid, which in turn dissociates to give hydrogen ions and hydrogen carbonate ions:

$$CO_2 + H_2O \rightleftharpoons H_2CO_3 \rightleftharpoons H^+ + HCO_3^-$$

Carbon dioxide is excreted via the lungs. The respiratory centre in the medulla oblongata increases the rate of breathing when the carbon dioxide concentration in the blood rises. Hydrogen ions are removed by the kidneys to maintain blood pH.

Kidney structure

The **urinary system** is made up of the:

- **kidneys**;
- **ureters**, taking urine from the kidneys;
- **bladder**, for the temporary storage of urine;
- **urethra**, taking urine from the bladder to the outside world.

Each human kidney is made up of around a million tubules, called **nephrons**, with their associated blood vessels. Different parts of the nephrons lie in the outer layer of the kidney, the **cortex**, and in the inner layer, the **medulla**. The kidney receives blood from the aorta via the **renal artery** and returns it to the vena cava via the **renal vein**. A longitudinal section (LS) of a kidney and the structure of a nephron are shown below.

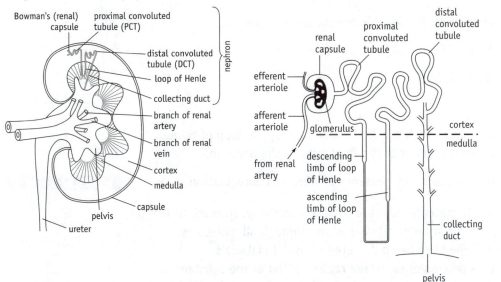

'proximal' means nearer to the starting point and 'distal' means further from the starting point.

An *afferent* blood vessel carries blood towards a structure; an *efferent* vessel carries it away.

A nephron begins as a concave **Bowman's capsule** or **renal capsule** in the cortex and leads to a highly coiled **proximal convoluted tubule (PCT)**. This leads on to a U-tube called the **loop of Henle** that lies in the medulla. The ascending limb of the loop of Henle leads to another coiled tubule in the cortex, the **distal convoluted tubule (DCT)**, and this leads to a **collecting duct**. A number of nephrons feed into the same collecting duct, which opens into the pelvis of the kidney.

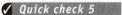 *Quick check 4*

Each Bowman's capsule has an associated **afferent arteriole** bringing blood from the renal artery to a network of capillaries called a **glomerulus**. The **efferent arteriole** draining this network is smaller in diameter than the afferent vessel. The efferent arteriole divides into capillaries that run close to the convoluted tubules and loop of Henle before joining as venules to take blood back to the renal vein.

✓ *Quick check 5*

? Quick check questions

1. Explain what is meant by homeostasis.
2. Explain what is meant by negative feedback in a living organism.
3. Why is excretion necessary in living organisms?
4. Draw a nephron and label its different regions.
5. Trace the route of a red blood cell from renal artery to renal vein.

Kidney function

The kidney is concerned both with **excretion** and with **homeostasis**. It:

● excretes urea and other waste solutes as urine;

● regulates the water content of the blood (**osmoregulation**);

● regulates the pH of the blood by excreting hydrogen ions if necessary.

These different functions are performed by different regions of the nephron, as outlined below and in the figure. Hydrogen ions are excreted by most parts of the nephron when the blood pH is too low.

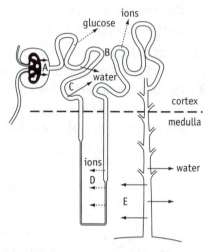

Cortex

A: Ultrafiltration from glomerular capillaries into the Bowman's capsule.

B: Selective reabsorption by the PCT and DCT.

C: Absorption of water by osmosis by the PCT.

Medulla

D: Active transport and diffusion of sodium and chloride ions by the loop of Henle to increase the solute concentration of tissue fluid outside the nephron.

E: Reabsorption of water by osmosis in the presence of ADH – osmoregulation.

✓ *Quick check 1*

It is not possible to filter waste material from the plasma selectively. Instead, in the process of **ultrafiltration**, the filtration membrane allows through all molecules below a certain size. To avoid the body being depleted of useful materials, ultrafiltration is followed by a process of **selective reabsorption** of the substances the body cannot afford to lose. At the same time most of the water lost from the blood during ultrafiltration is returned to the blood and the hormonally controlled mechanism of osmoregulation takes place.

Ultrafiltration

● The afferent arteriole supplying the capillaries of the glomerulus is wider than the efferent arteriole. This, and the short distance of the kidneys from the aorta, provides the necessary hydrostatic **filtration pressure**.

● The filtration membrane is the basement membrane of glycoproteins surrounding the endothelium of the capillaries. The inner layer of the Bowman's capsule consists of highly adapted **podocytes** (see below), which provide support for the membrane with minimum interruption to filtration.

❶ Crytalloids are substances that form a true solution. Colloids have particles of about 100–10 000 nm in diameter, which remain dispersed.

● Water and crystalloid solutes, including urea, are filtered from the blood into the Bowman's capsule. Cells, platelets and colloidal solutes, including the blood proteins with a relative molecular mass greater than 69 000, do *not* pass through the undamaged filtration membrane.

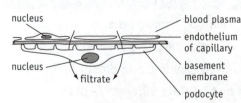

✓ *Quick check 2*

Selective reabsorption

- Glucose, amino acids, vitamins and some ions are reabsorbed by the PCT.

- Sodium ions are actively transported out of the PCT cells through the plasma membranes next to blood capillaries, so sodium diffuses into the cells from the filtrate via **cotransporter** proteins, bringing glucose into the cells with the sodium ions.

- The PCT cells have microvilli to give a large surface area of contact with the filtrate and many mitochondria to provide ATP energy for active transport.

- Ions are absorbed from the filtrate by the DCT cells, which, although smaller, have the same adaptations as the PCT cells.

✓ *Quick check 3*

Passive processes

❂ Remind yourself of the transport of substances in and out of cells from your AS course.

- 65% of the water filtered from the blood by ultrafiltration is reabsorbed at the PCT by osmosis, because the filtrate has a *higher* **water potential** than the blood plasma it was filtered from.

- Some urea diffuses out of the filtrate at the PCT.

- In the presence of **antidiuretic hormone (ADH)** from the **posterior pituitary** gland, the collecting duct wall is permeable to water, which passes out of the filtrate by osmosis. In the absence of ADH the collecting duct wall is impermeable to water. This is osmoregulation.

The countercurrent mechanism

❂ Fluid flows in opposite directions in the two limbs of the loop of Henle: hence 'countercurrent'

- Sodium and chloride ions move by active transport and diffusion out of the filtrate as it moves along the **ascending limb** of the **loop of Henle**.

- This decreases the water potential of the tissue fluid in the tips of the pyramids of the medulla of the kidney, through which the collecting ducts pass.

- The ions pass back into the **descending limb** and are recycled. This allows them to be concentrated in the medulla.

- The low water potential outside the collecting duct allows water to leave the collecting duct by osmosis, provided ADH is present.

✓ *Quick check 4,5*

❓ Quick check questions

1 List the processes taking place in (i) the cortex and (ii) the medulla of the kidney.

2 What is the role of a podocyte in ultrafiltration?

3 Explain why the process of selective reabsorption is necessary in the kidney.

4 Explain the return, at the PCT, of much of the water filtered from the blood at the Bowman's capsule.

5 Explain how the medulla of the kidney is able to perform osmoregulation.

Communication

All living organisms must respond to their environment. Mammals need a communication system between **receptors** and **effectors** (see page 62), which may be far removed from one another, so that they can respond adaptively to changes in both internal and external environments. The communication system has two components.

- Nerve cells or **neurones** make up a **nervous system** (see page 68). In this system, a change in the internal or external environment is detected by a receptor that generates electrical nerve impulses. These pass along specific nerve pathways via the central nervous system to the effector.

- Chemical messengers or **hormones** and the glands that secrete them make up the **endocrine system** (see page 74). In this system, a change in the internal or external environment results in either the secretion of a chemical or stopping its secretion. On secretion, the chemical travels in the lymph and blood all over the body but affects only certain target tissues (effectors).

The differences between these two communication systems mean that they tend to be used for different purposes, and it is useful to compare the two.

	nervous system	endocrine system
information transfer	nerve impulses: all-or-nothing changes in potential difference on the surface of neurones	hormones: molecules travelling in the lymph and blood plasma
speed of transfer	fast: up to 100 m sec^{-1}	slow: move with the blood stream
delivery	the nervous system is 'hard wired': nerve impulses in a particular neurone travel to specific destinations	hormones move in body fluids all over the body
effect	tend to produce a rapid and short-lived response	the response is generally slower and hormones are often used in long-term responses such as growth and development

✓ *Quick check 1,2*

Sensory receptors

A **sensory receptor** is a cell in which a change in the internal or external environment produces a nerve impulse or action potential (see page 70). Some receptors are scattered in appropriate body tissues, such as the skin or the lining of the airways. Others are concentrated into organs such as the eye or ear. All receptor cells are **transducers** in which one form of energy, the stimulus energy, is converted to another, electrical.

An example of a non-biological transducer is a microphone, which converts sound energy to electrical energy.

Mammalian receptors

Receptors in mammals are adapted to respond to different aspects of the environment, such as:

- light intensity and wavelength;
- sound;
- touch and pressure;
- temperature;
- chemicals.

type of receptor	type of stimulus energy	stimulus
photoreceptor	electromagnetic	light intensity and wavelength
electroreceptor	electromagnetic	electricity
mechanoreceptor	mechanical	sound, touch, pressure, gravity
thermoreceptor	thermal	temperature change
chemoreceptor	chemical	humidity, smell, taste, water potential, ion concentration

✔ *Quick check 3, 4*

Each type of receptor must have the means of absorbing its particular stimulus energy. For example, a **rod cell** in the retina, which can respond to the presence of light but cannot detect its wavelength, absorbs light by means of a pigment, **rhodopsin**. (The pigment is slightly purple in white light and is also called **visual purple**.) When light energy is absorbed by rhodopsin, the pigment molecule changes shape. This in turn leads to a change in permeability of the rod cell membrane. One photon of light falling onto a rod cell is sufficient to trigger a response.

Compare the changing shape of rhodopsin with shape change in other molecules you have met, such as enzymes.

Similarly, the three different types of **cone cell** of the retina have slightly different light-absorbing pigments (**opsins**) which give them different sensitivities to different wavelengths of light. The three types of cone absorb in the blue, green or red wavelengths of the spectrum and together give colour vision.

A **Pacinian corpuscle** is a pressure or vibration receptor in human skin. It consists of a sensory nerve ending enclosed in concentric layers of flattened cells. When pressure is applied to these layers the permeability of the nerve ending to sodium ions is increased, and when the pressure is great enough a nerve impulse is set up.

✔ *Quick check 5*

You will learn more about the nervous system and sense organs if you choose Mammalian Physiology and Behaviour as your optional module.

? *Quick check questions*

1 Explain why mammals need a communication system to respond to environmental changes.

2 List the main differences between communication by means of the nervous system and the endocrine system.

3 Explain why sensory receptors are described as transducers.

4 List the different types of receptor found in mammals.

5 Describe how a named receptor is able to act as a transducer.

Nerve cells (neurones)

Cells that are specialised to transmit the electrical signals of nerve impulses very rapidly from one part of the body to another are called **nerve cells** or **neurones**. There are three basic types of neurone in the mammalian nervous system:

- **sensory neurones**, which transmit nerve impulses from a receptor to the central nervous system (brain or spinal cord);
- **motor neurones**, which transmit nerve impulses from the central nervous system to an effector such as a muscle or a gland;
- **interneurones** (intermediate or relay neurones), which transmit nerve impulses between other neurones, e.g. between a sensory and a motor neurone.

Structure of neurones

A mammalian **motor** neurone has:

- a cell body or **centron**, with a large nucleus and large quantities of rough endoplasmic reticulum and Golgi bodies;
- many short **dendrites** that carry nerve impulses *towards* the cell body;
- a very long process called the **axon** that carries impulses *away* from the cell body to an effector.

A mammalian **sensory** neurone has long processes on either side of the centron. The process carrying nerve impulses *towards* the cell body, from a receptor, is the **dendron**. An axon carries nerve impulses *away* from the cell body, this time to the central nervous system.

✓ *Quick check 1*

Interneurones have many shorter processes around their cell bodies, carrying impulses from sensory neurones to motor neurones.

The three types of neurone can be seen below.

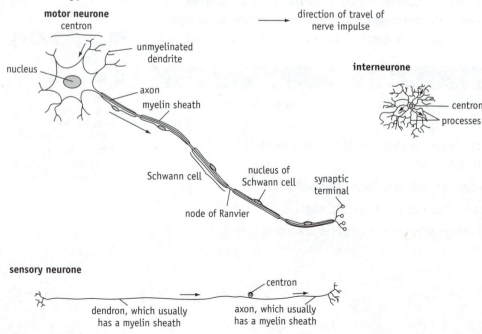

motor neurone
centron

direction of travel of nerve impulse

nucleus

unmyelinated dendrite

interneurone

axon
myelin sheath

centron
processes

Schwann cell

nucleus of Schwann cell

synaptic terminal

node of Ranvier

sensory neurone

centron

dendron, which usually has a myelin sheath

axon, which usually has a myelin sheath

Myelin

Axons and dendrons have cells wrapped round them that nourish and protect them. These cells are **Schwann cells**. In mammals, the Schwann cells wrap themselves many times around an axon or dendron, producing multiple layers of plasma membrane called **myelin**. The gaps between individual Schwann cells are called **nodes of Ranvier**. A myelin sheath helps to speed up the transmission of a nerve impulse from about 4 m sec^{-1} to 100 m sec^{-1} (see page 71).

✓ *Quick check 2*

The reflex arc

The **reflex arc** is the simple pathway by which nerve impulses are passed from a receptor to an effector via the central nervous system. The arc usually, but not always, involves an interneurone between the sensory and motor neurones. Neurones do not make direct contact, but are separated by **synaptic clefts** at **synapses** (see page 72). The narrow processes of neurones, each surrounded by its myelin sheath, are clustered into bundles or **nerves** protected by layers of connective tissue. Nerves may be:

- **cranial nerves**, taking impulses to and from the brain;
- **spinal nerves**, taking impulses to and from the spinal cord.

✓ *Quick check 3*

The figure shows a spinal reflex arc.

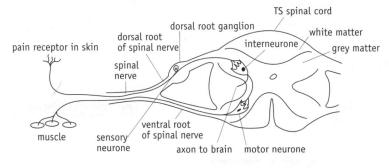

The reflex arc is a rapid pathway between receptor and effector. It enables a particular response to result rapidly from a particular stimulus, without the need to involve the decision-making processes of the brain. This is particularly valuable for protecting the body from damage. Examples are:

- removing a hand from a hot surface;
- blinking as an object approaches the eye.

✓ *Quick check 4, 5*

Interneurones do not only complete the reflex arc, they also transmit the nerve impulse to the brain. Depending on the distances involved, the brain may therefore be 'aware' of the activity of the reflex arc as the response happens or slightly after it.

The brain can modify the activity of the reflex arc by inhibiting the motor neurone so that it does **not** respond so easily to the incoming impulses from the sensory neurone.

? Quick check questions

1. List the similarities and differences in structure and function between sensory and motor neurones.

2. Describe the myelin sheath of a neurone.

3. Explain what is meant by the term nerve.

4. Explain how a reflex arc helps to protect the body from unnecessary damage.

5. Distinguish between cranial and spinal reflexes and give one example of each.

Transmission of a nerve impulse

A **nerve impulse** is an electrical event. Each impulse is a brief change in the potential difference across the plasma membrane of the neurone that passes from one end of the cell to the other. The potential difference across the plasma membrane when the cell is at rest is called the **resting potential**. The brief change in polarity that is a nerve impulse is called the **action potential**.

> ❯ Be careful to talk about a potential difference *across* a membrane.

The resting potential

- When **not** conducting a nerve impulse there is a **potential difference (p.d.)** across the plasma membrane such that the inside has a negative charge of –70 mV compared with the outside (see opposite).
- Sodium–potassium pumps, in the plasma membrane, pump K^+ ions into the cell and Na^+ ions out.
- Three Na^+ ions are pumped out for every two K^+ ions pumped in.
- The axon contains organic anions to which the membrane is impermeable.
- This, and a slight loss of K^+ ions through the permeable membrane, accounts for the resting potential.
- The membrane is effectively impermeable to Na^+ ions.

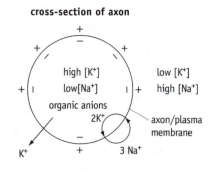
cross-section of axon

✓ *Quick check 1*

The action potential

- When the membrane is depolarised and the p.d. across the membrane reaches about –40 mV (the **threshold to fire**), the sodium ion channel proteins in the membrane suddenly open.
- As Na^+ ions are in high concentration outside the cell and low concentration inside, Na^+ ions flood into the cell, making it positive inside with respect to outside and changing the p.d. to +40 mV inside the cell (see opposite).
- The p.d. achieved in a particular neurone is always the same.
- The sodium ion channels now close. Potassium ion channels open and potassium ions diffuse out of the cell. The p.d. becomes negative inside the cell again, falling to about –75 mV. The membrane is **hyperpolarised**.
- Most potassium ion channels close and the sodium–potassium pump restores the resting potential.
- This sequence of events is an **action potential**. Transmission of an action potential along an axon forms a nerve impulse.

The changes in potential difference during an action potential are shown opposite.

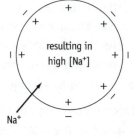
cross-section of axon

✓ *Quick check 2,3*

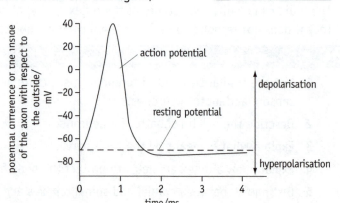

Movement of an action potential

The action potential travels along the plasma membrane. This is caused by the movement of sodium ions, both inside and outside the axon, towards a negatively charged region. This in turn begins to depolarise the membrane ahead of the action potential. When the depolarisation reaches the threshold to fire, the sodium ion channels open, giving an action potential.

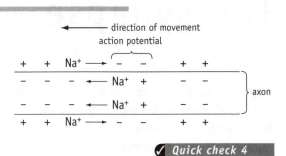

✓ *Quick check 4*

In a non-myelinated mammalian neurone the action potential is transmitted along the axon membrane at about 4 m sec^{-1}; the actual speed depends on the diameter of the axon. A **myelinated** cell is able to conduct nerve impulses much faster by effectively 'jumping' them from one gap, a **node of Ranvier,** in the insulating myelin to the next. The nodes are 1–2 mm apart. Local circuits set up by the presence of the action potential at one node depolarise the membrane at the next node to the threshold to fire. The action potential has 'jumped' ahead of the slower wave of depolarisation passing along the axon membrane.

✓ *Quick check 5*

Refractory period

When the membrane is hyperpolarised, depolarisation will not reach the threshold to fire. The membrane can only respond to depolarisation when its resting potential has been restored. This means that there is an unresponsive period: the **refractory period**. This imposes a gap in space and time between one action potential and another, and results in the impulse travelling in one direction.

Information transfer

> Remember that the size of an action potential in a particular cell is always the same.

The **type** of stimulus is recognised by the source of the information. Neurones from particular receptors pass to particular positions in the brain.

The **strength** of the stimulus is recognised by the number of cells responding to the stimulus and by the frequency of action potentials on those cells.

When there is sufficient stimulus to depolarise the membrane to the threshold to fire, an action potential follows. When the depolarisation does not reach the threshold to fire, no action potential results. This means that an action potential is **all-or-nothing**. Hence, no information about the strength of the stimulus can be transmitted by differences in size of action potentials. Instead, a strong stimulus results in many action potentials following one another as closely as possible.

? *Quick check questions*

1 Explain what gives rise to the resting potential of a neurone.

2 List the events that result in an action potential.

3 Distinguish between depolarisation and hyperpolarisation.

4 Explain how an action potential moves along a non-myelinated axon.

5 Explain how the presence of a myelin sheath speeds up the movement of an action potential.

The synapse

A **synapse** is where one neurone makes functional contact with another neurone or with an effector. A neurone may have up to 10 000 synapses with 1000 other cells. Most synapses are between the axon terminals of the stimulating neurone and the dendrites of the stimulated neurone. Typically a gap, the **synaptic cleft**, of about 15 nm is found between the two cells. Most synapses use **neurotransmitters** such as **acetylcholine (ACh)** to cross this gap, but some synapses are purely electrical. Synapses in which the neurotransmitter is ACh are said to be **cholinergic**. The structure of a typical synapse is shown below.

✓ *Quick check 1*

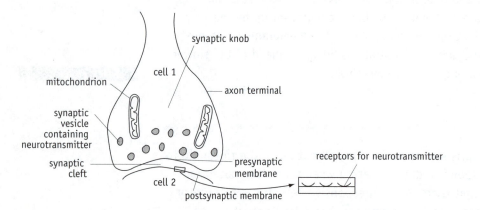

✓ *Quick check 2*

Transmission across a cholinergic synapse

The events taking place during transmission of a nerve impulse across a synapse are summarised below. Relate each numbered event to the figure opposite.

1 An action potential arrives at the axon terminal of cell 1.

2 Calcium ion channels open and calcium ions (Ca^{2+}) diffuse into the axon terminal (**synaptic knob**).

3 This causes the **synaptic vesicles**, which contain ACh made from choline and acetylcoenzyme A (see page 8), to move to the **presynaptic membrane**.

4 The vesicles fuse with the presynaptic membrane and release ACh into the synaptic cleft.

5 ACh diffuses across the synaptic cleft and attaches to receptors on the **postsynaptic membrane** (cell 2).

6 This causes the sodium channels to open. Sodium ions enter cell 2, depolarising the membrane and starting a new action potential.

7 The enzyme **acetylcholinesterase** (attached to the postsynaptic membrane) breaks down ACh. Much of the choline is absorbed through the presynaptic membrane and remade into ACh.

❶ *pre–* = before; *post–* = after

❶ Enzymes have the ending *–ase*. In this case the enzyme breaks down an organic salt or *ester*.

✓ *Quick check 3*

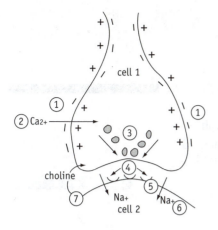

Roles of synapses

Synapses:

- allow transmission in **one** direction only, from presynaptic to postsynaptic membrane;

- may be **excitatory** or **inhibitory**, providing flexibility of response;

- allow **spatial** and **temporal summation**;

- allow **facilitation** in which the arrival of each action potential leaves the postsynaptic membrane more responsive to the next.

The neurotransmitter of an excitatory synapse depolarises the postsynaptic membrane as described above, but that of an inhibitory synapse **hyperpolarises** the postsynaptic membrane, preventing any depolarisation from reaching the **threshold to fire** (see page 70).

When two action potentials arrive at two excitatory synapses on the same neurone at the same time the depolarisation of the postsynaptic membrane will be greater, but when action potentials arrive at an excitatory and an inhibitory synapse at the same time their effects cancel out and no depolarisation of the postsynaptic membrane occurs. This integration mechanism is **spatial summation**.

When two action potentials arrive very closely after one another the depolarising effect is greater: this is **temporal summation**.

✓ *Quick check 4, 5*

? Quick check questions

1 Explain what is meant by the terms synapse and neurotransmitter.

2 Draw and label the structure of a typical synapse.

3 List the sequence of events as a nerve impulse is transmitted across a cholinergic synapse.

4 List the roles played by synapses in the nervous system.

5 Explain what is meant by the terms summation and facilitation in the context of a synapse.

Endocrine glands

An **endocrine gland** secretes its product directly into the blood and/or lymph (see below). Together, the endocrine glands form the **endocrine system** (see page 66). An **exocrine gland** has a duct to take the gland's secretion to its site of action, e.g. salivary glands with ducts to the buccal cavity.

✓ Quick check 1

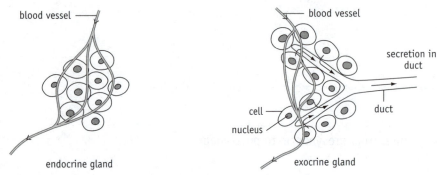

The product of an endocrine gland is a **hormone**. A hormone:

- is a molecule that acts as chemical messenger;
- travels in the blood and lymph all over the body;
- causes a specific response in specific target cells or tissues;
- may be a steroid, such as oestrogen, progesterone and testosterone;
- may be a polypeptide, such as insulin, glucagon, follicle-stimulating hormone (FSH) and luteinising hormone (LH);
- may be an amino acid, such as thyroxine;
- may be a catecholamine, such as adrenaline.

❶ A catecholamine is a derivative of the amino acid, tyrosine.

✓ Quick check 2

Examples of endocrine glands and some of their hormones and target tissues are shown below.

endocrine gland	hormone	target tissue
anterior pituitary	*FSH, LH*	*ovary or testis*
	thyroid-stimulating hormone (TSH)	*thyroid*
	somatomedin	*skeleton, muscle*
posterior pituitary	**antidiuretic hormone (ADH)**	**kidney nephron collecting duct**
thyroid	*thyroxine*	*all aerobically respiring cells*
islets of Langerhans	**insulin and glucagon**	**liver, skeletal muscle and fat cells**
adrenal medulla	adrenaline	heart, smooth muscle, liver
adrenal cortex	corticoids	kidney, gut
follicle cells of ovary	*oestrogen*	*uterus, breasts, skeleton*
corpus luteum of ovary	*progesterone*	*uterus, breasts*
interstitial cells of testis	*testosterone*	*skeleton, muscles*

❶ For this module you only need to know the ones in **bold**. The examples in *italics* will be needed if you study Growth, Development and Reproduction as your optional module.

Hormones act in two different ways when they reach a target tissue.

- Polypeptide hormones and catecholamines (adrenaline) do **not** enter the cells of the target tissue. They bind with receptors on the plasma membrane of the cell.

This triggers a series of events within the plasma membrane and then within the cell, where enzymes are activated. This type of response involves **secondary messengers** within the cell.

- Steroids and thyroxine pass through the plasma membrane and bind with receptors in the nucleus. The hormone–receptor **complex** binds with DNA and regulates transcription.

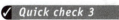
✓ *Quick check 3*

The endocrine pancreas: the islets of Langerhans

The islets of Langerhans are patches of endocrine tissue scattered throughout the exocrine tissue of the pancreas (see below). Remember, the exocrine cells of the pancreas secrete digestive enzymes that travel via a duct to the duodenum. The islets of Langerhans form about 15% of the pancreas.

Each islet has a mixture of several different types of cell, including:

- α cells, which secrete the hormone **glucagon**;

- β cells, which secrete the hormone **insulin**.

Both α and β cells have receptors in their plasma membranes that enable them to detect changing concentrations of plasma glucose.

✓ *Quick check 4*

> You will learn more about the exocrine pancreas if you study Mammalian Physiology and Behaviour as your optional module.

✓ *Quick check 5*

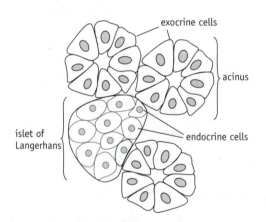

Quick check questions

1 Distinguish between endocrine and exocrine glands.

2 Explain what is meant by the term hormone.

3 Describe the two different methods of action of hormones on target tissues.

4 Explain why the name 'islets' is applied to the endocrine pancreas.

5 Describe the different roles of exocrine and endocrine tissue in the pancreas.

Regulation of plasma glucose concentration

Plasma glucose concentration is regulated by **negative feedback** (see page 62) as shown in below.

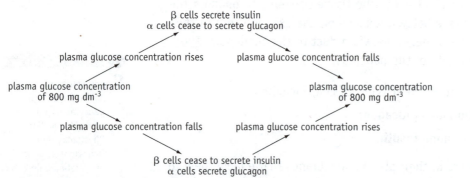

Suppose that the plasma glucose concentration **rises** because glucose is absorbed from food in the gut or released from the liver. Then:

- both α and β cells detect the rising concentration of glucose;
- α cells cease to secrete glucagon but β cells secrete **insulin**;
- insulin binds to the receptors in the plasma membranes of muscle, fat and liver cells;
- glucose uptake from the blood by these cells is increased;
- use of glucose in respiration by the cells is increased;
- liver cells convert glucose to glycogen and stop the reverse reaction.

Conversely, when plasma glucose **falls** because it is used rapidly by cells or is lacking in the food eaten:

- both α and β cells detect the falling concentration of glucose;
- β cells cease to secrete insulin;
- α cells secrete **glucagon**;
- target cells take up less glucose;
- the rate of use of glucose decreases and fats or amino acids are used instead in respiration;
- glycogen in the liver is converted to glucose and released into the blood.

Insulin secretion

A β cell has a resting potential across its membrane that is kept negative inside the cell by ATP-sensitive potassium ion (K^+) channels. When plasma glucose concentration rises, glucose enters the cell and is respired, allowing ATP production. The increased ATP concentration in the cell closes the K^+ channels. The membrane depolarises and calcium ions rush into the cell. Insulin is released by exocytosis. In this way, the K^+ channels control the **set point** for plasma glucose homeostasis (see page 62).

> A healthy human's blood normally contains 80–90 mg per 100 cm^3 or 800–900 mg dm^{-3} glucose.

> Revise blood plasma from your AS course.

> You will learn more about the liver's role in carbohydrate metabolism if you study Mammalian Physiology and Behaviour as your optional module.

> ✓ Quick check 1, 2

> Compare this mechanism with release of ACh at a synapse.

> ✓ Quick check 3

Diabetes mellitus

Diabetes mellitus arises when no insulin, or not enough insulin, is produced, or when the available insulin does not function correctly.

Insulin-dependent (type 1) diabetes occurs when there is a lack of insulin because of the destruction of all or most of the β cells of the pancreas. Regular injections of insulin are needed for survival. The destruction of the β cells is an autoimmune response. The uptake of glucose by brain cells and red blood cells is not dependent on the presence of insulin, but even in the presence of abundant glucose the sugar is not taken up by muscles in the absence of insulin.

Non-insulin-dependent (type 2) diabetes occurs when the body does not produce enough insulin and that which is produced becomes less effective on the target cells. This type of diabetes can usually be controlled by diet, but sometimes insulin injections are needed.

✓ *Quick check 4*

Human insulin from genetically modified bacteria

There are differences in the amino acid sequence between human insulin and the insulin extracted from cows, sheep and pigs. Most insulin available for injection is from pigs or human sequence insulin prepared either by modifying pig insulin or produced by bacteria into which the gene for human insulin has been added.

▶ Revise the synthesis of human insulin by bacteria from your AS course.

Manufacture of human insulin by *Escherichia coli* was one of the early successes in biosynthesis of human proteins. The **advantages** of using human insulin produced by genetically modified bacteria are as follows.

- There is less likelihood of an immune response producing anti-insulin antibodies.
- There is a more rapid response and a shorter duration of response.
- Patients who have become tolerant of pig insulin (requiring larger doses to have an effect) respond better to human insulin.
- Some patients do not like the idea of using insulin from animals, e.g. for religious reasons or because of the BSE outbreak.
- Extraction and purification of insulin from animal pancreases is expensive.

▶ Revise the immune response from your AS course.

One reported **disadvantage** of using human insulin is that some patients no longer experience any warning signs of low plasma glucose concentration: hypoglycaemia.

✓ *Quick check 5*

? *Quick check questions*

1 Identify the receptors and effectors involved in the homeostasis of plasma glucose.
2 Describe how plasma glucose concentration is lowered.
3 Describe how a β cell releases insulin.
4 Distinguish between type 1 and type 2 diabetes mellitus.
5 List the advantages of treating diabetics with insulin produced by bacteria.

Communication in flowering plants

Like mammals, flowering plants need a communication system to respond adaptively to changes in their internal and external environments. A seed must germinate in suitable conditions and the radicle and plumule grow down and up, respectively. The plant must then grow adaptively with respect to light intensity and variations in water content of soil. A plant growing in a region with seasons must respond to those seasons and flower at the right time.

Tropisms

Many of the responses of plants are by means of **growth**. A growth response is called a **tropism** when the growth is in a direction determined by the stimulus. A response towards the stimulus is said to be **positive**, and one away from the stimulus **negative**. The table shows some tropisms.

✓ *Quick check 1*

❶ Avoid using the term 'hormone' for plant growth substances.

stimulus	growth response
light	phototropism
gravity	geotropism
touch	thigmotropism

The communication system of flowering plants involves:

sensor (receptor) ⟶ **plant growth substances** ⟶ **effector**
(or plant growth regulators)

✓ *Quick check 2*

Plant receptors

If a flowering plant is to respond to a change in its internal or external environment it must have a receptor that is sensitive to that particular change.

For example, a plant that can respond to light by growing towards the source of light **(positive phototropism)** needs a photoreceptor molecule that is affected by light. Much work has been done on the phototropic response of cereal **coleoptiles**, in which the tip of the coleoptile is the receptor region of the plant. A cereal coleoptile responds most to wavelengths in the blue region of the spectrum around 450 nm. It is not yet certain what the photoreceptor is, but two molecules in the coleoptile tip have absorption spectra with maxima in that region: **carotene** and **riboflavin**. It is likely that riboflavin is the photoreceptor for phototropism.

❶ A coleoptile is a protective sheath surrounding the first leaf of a germinating grass or cereal.

The photoreceptor controlling germination of light-sensitive seeds and flowering that is dependent on day-length is **phytochrome**. This pigment is a small protein that has two interconvertible forms. P_r absorbs red light with a wavelength of 660 nm, and P_{fr} absorbs far-red light with a wavelength of 730 nm. When P_r absorbs red light it is converted to P_{fr}. When P_{fr} absorbs far-red light, or is kept in the dark, it is converted into P_r.

$$P_r \quad \overset{\text{red light}}{\underset{\substack{\text{far-red light} \\ \text{or darkness}}}{\rightleftharpoons}} \quad P_{fr}$$

P_{fr} is the active form of phytochrome. Sunlight increases the amount of P_{fr} in light-sensitive seeds and this stimulates germination. P_{fr} inhibits flowering in so-called short day plants. Only when the nights are long enough for enough P_{fr} to be broken down to P_r can flowering occur.

✓ Quick check 3

Plant growth substances

There are five types of plant growth substance.

- **Auxins** stimulate cell elongation in stems, produce apical dominance and stimulate the growth of adventitious roots in stem cuttings. The first auxin to be discovered was **indoleacetic acid (IAA)**, which is involved in phototropism.
- **Gibberellins** promote stem elongation and stimulate seed germination.
- **Cytokinins** promote cell division and delay senescence.
- **Abscisic acid (ABA)** inhibits seed germination, thereby promoting seed dormancy, closes stomata in conditions of water stress and helps plants adapt to cold conditions.
- **Ethene** ripens fruit and is involved in responses to stress.

✓ Quick check 4

Plant effectors

Tropisms are growth responses and the effector is the **region of cell elongation** just behind a shoot, coleoptile or root tip. In positive phototropism of a coleoptile, the cells on the shaded side elongate more than those on the illuminated side. Hence the coleoptile bends towards the source of light.

The effector in cereal seed germination is a layer of cells, the **aleurone layer**, which secretes the enzymes needed to start off the process of germination (see page 80). In flowering, the effector is a growing bud in which the genes responsible for producing leaf primordia are switched **off** and the genes responsible for producing flower primordia are switched **on**.

✓ Quick check 5

? Quick check questions

1 Explain what is meant by a tropism.
2 State the essential components of a communication system in a flowering plant.
3 State an example of a receptor in a flowering plant.
4 List the types of plant growth substances.
5 State one example of an effector in a flowering plant.

Roles of plant growth substances

Auxins and apical dominance

In many plants, side shoots do not grow if the main shoot is growing. However, if the bud at the tip of the main shoot (the **terminal** or **apical bud**) is removed, then the **axillary** or **lateral buds** in the axils of leaves lower down the plant start growing. This effect may be caused by **auxins** produced by the apical bud inhibiting growth of lateral buds. Removing the apical bud removes the source of the inhibitory auxin. The experimental evidence for this effect of auxin is contradictory and cytokinins and abscisic acid may also be involved.

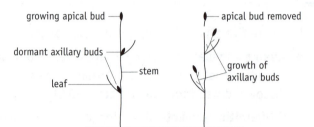

As the plant grows taller, the lateral buds furthest away from the tip may start growing. Because the auxin is diluted as it passes down the stem from the apical bud, lateral buds at the base of the stem may no longer be inhibited because the concentration of auxin reaching them is below the threshold value for inhibition. Once active, the lateral buds in their turn produce auxins and cytokinins. **Apical dominance** gives a plant a shape that allows all parts access to light.

✓ *Quick check 1*

Gibberellins and stem elongation

Most plants produce **gibberellins** to regulate **stem elongation**. The effect is seen most obviously when genetically dwarf plants are treated with **gibberellic acid (GA)**: the stems elongate considerably. Similarly, plants with rosettes of leaves separated by very small internodes elongate the internodes when treated with GA. This is seen naturally when a compact lettuce plant 'bolts' before flowering.

Gibberellin and germination of cereals

In a grain of wheat or barley (which are monocotyledons), the embryo with its single cotyledon is surrounded by the food store of the endosperm. The food store is mainly insoluble starch, which must be hydrolysed to sugars before it can be absorbed by the cotyledon and transported to the embryo to be used in growth. When germination is initiated, the following sequence of events occurs:

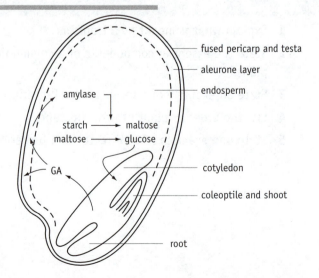

- GA secreted by the cotyledon diffuses across the endosperm to the aleurone layer;

- GA activates the gene coding for amylase;

- amylase is synthesised by the aleurone layer and diffuses into the endosperm;
- amylase hydrolyses starch to maltose and this is hydrolysed to glucose which can diffuse into the embryo.

✓ Quick check 2

Abscisic acid and leaf fall

Abscisic acid (ABA) is a plant growth substance that stimulates the production of **ethene**. Ethene stimulates the fall of leaves or **abscission**. The following sequence of events leads to leaf fall:

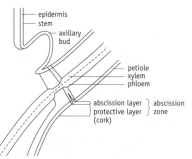

- cells in a layer at the base of the petiole separate from one another by the breakdown of the middle lamellae of their cell walls;
- a protective layer of cells with wax in their walls forms on each side of this **abscission layer** to prevent infection and water loss, in woody species this layer is cork with suberised cell walls;
- the vascular tissue is sealed;
- the leaf is broken off by a mechanical force such as the wind.

✓ Quick check 3

Abscisic acid and stomata

◐ Revise transpiration from your AS course.

ABA can help a plant survive stress. When water is in short supply, ABA is released from the chloroplasts of mesophyll cells. ABA concentrations in wilted tomato plant leaves have been found to be 20 times that in unwilted leaves. When ABA reaches the guard cells of a stoma, they lose their turgidity and the stoma closes.

When a guard cell takes up water by osmosis, the thick wall adjacent to the stoma resists stretching and becomes more curved. In this way the pore opens. Uptake of water occurs because the water potential of the cell is lowered by addition of potassium ions (K^+). An ATP-dependent (see page 4) proton pump transports H^+ ions out of the cell. The reduced H^+ concentration inside the cells causes potassium ion channels in the plasma membrane to open and K^+ ions move down an electrochemical gradient into the cell. Guard cells have ABA receptors in their plasma membranes. Binding ABA with these receptors may inhibit the proton pump.

Spraying ABA onto crops as an anti-transpirant is not a commercial proposition as the effects are short-lived. A methyl ester of ABA can be used because it is not readily broken down by plants.

✓ Quick check 4, 5

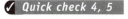

? Quick check questions

1. Explain what is meant by apical dominance.
2. List the roles of gibberellic acid in a flowering plant.
3. Describe the events leading to leaf abscission.
4. Describe the effect of ABA on a turgid guard cell.
5. Suggest circumstances in which it might be necessary to spray a crop with an anti-transpirant.

Exam-style questions

1 A radioactive isotope of sodium has the same chemical properties as the commonly occurring isotope and can be used to trace the movement of sodium ions in biological systems.

A squid giant axon was placed in a dish fitted with a drain, so that the solution bathing the axon could be easily removed and replaced without disturbing or stimulating the axon.

When the solution had been removed, the radioactivity of the axon could be measured with a Geiger counter. The count rate registered by the Geiger counter was directly proportional to the number of radioactive sodium ions within the axon.

Once each hour the axon was bathed in a saline solution containing some radioactive sodium ions. While in this radioactively labelled saline, the axon was electrically stimulated. The saline bathing the axon was then replaced by saline containing only the normal isotope of sodium. The radioactivity of the axon was measured every 15 minutes, until the time came for the next stimulus to be administered.

For the first 2 hours, the axon was placed in a solution containing radioactive sodium ions, but it was not stimulated electrically.

The results of the investigation are shown in the figure.

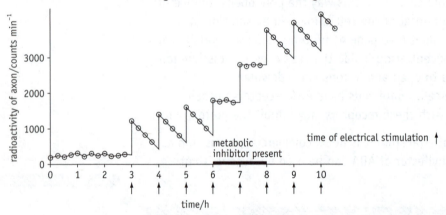

During the period between the sixth and the eight hours, the axon was prevented from synthesising ATP by the use of a metabolic inhibitor. This inhibitor was added to all bathing solutions during this period.

With reference to the figure:

a state the effect of an electrical stimulation on the radioactivity of the axon; [2]

b explain why an electrical stimulation produced this effect; [3]

c describe the changes in the radioactivity of the axon between the second and third stimulus and between the fifth and sixth stimulus; [2]

d explain the effect of the inhibitor on the radioactivity of the axon. [4]

2 a State **three** differences between nervous and hormonal control in mammals. [3]

The embryos were removed from a number of dry, freshly harvested barley seeds.

These seeds were then divided into two similar batches. One batch was soaked overnight in a small volume of water. The other batch was soaked in the same volume of water to which a number of embryos, detached from germinating seeds, had been added.

See page 80 to remind yourself about barley seeds.

Each embryo-free seed was then placed on the surface of agar jelly in a Petri dish and covered with a small drop of water. The detached embryos were treated in the same way.

The agar contained starch.

After 24 hours, the agar in each dish was covered in potassium iodide solution. It went black where it contained starch. The agar under the seeds was free of starch and the area of these starch-free zones was estimated using graph paper. The results are shown in the table.

area of agar free of starch / mm^2		
seeds soaked in water	seeds soaked in the presence of embryos	embryos
8	16	6
12	17	4
9	18	6
9	16	7
11	15	3
mean =	mean =	mean = 5

With reference to the table:

b complete the table by entering the mean starch-free areas for the two batches of seeds; [2]

c explain these results. [5]

Answers to quick check Questions

Many of the quick check questions are designed to test your understanding and recall of material in the spreads. You will find many of the answers quite easily. However, some are more difficult and require you to put several ideas together. Answers to selected quick check questions are given here.

ATP (page 4)

1   **ii**

5   Changes in the shape of a binding site allow reactants to be brought close together to react. Another change allows the product to be released.

Respiration (page 6)

5

used per glucose	produced per glucose	net gain per glucose
−2 ATP	+4 ATP	+2 ATP

Aerobic respiration (page 8)

3

	NAD_{red}	FAD_{red}
glycolysis	2	
link reaction	2	
Krebs	6	2
total	10	2

5   Remember that each NAD_{red} yields 2.5 ATP and each FAD_{red} 1.5 ATP.

	ATP
NAD from glycolysis	$2 \times 2.5 = 5$
NAD from link reaction	$2 \times 2.5 = 5$
ATP from Krebs	2
NAD from Krebs	$6 \times 2.5 = 15$
FAD from Krebs	$2 \times 1.5 = 3$
total	30

Respiratory quotients (page 12)

2 $102/145 = 0.7$

Photosynthesis (page 18)

1   The molecule (RuBP) which combines with carbon dioxide. It 'fixes' carbon dioxide.

3   ATP supplies energy for the reduction of GP to triose and for the regeneration of RuBP. NAD_{red} supplies hydrogen needed for the reduction of GP to triose.

5   10 molecules (30 carbons) to 6 RuBP; 2 molecules (6 carbons) to 1 hexose or to fatty acids or amino acids.

Limiting factors (page 22)

2   Any three from: light intensity; temperature; water supply; stomatal opening; availability of chlorophyll, carriers and enzymes.

3   See the second graph on page 22. The added curve should plateau in the same way *below* the others.

4   Light intensity is the limiting factor.

Populations and interactions (page 26)

2   Lack of lactose; lack of oxygen; increased lactic acid; decreased pH.

3   See the graph on page 27. The curve will be the same as that showing carrying capacity.

Predator–prey interactions and competition (page 28)

1   The peaks in hare numbers come *before* peaks in lynx numbers.

2   Troughs in prey species lead to decreased numbers of owls. Peaks in prey numbers in autumn and early winter stabilise existing numbers of owls. (The owls do not increase in numbers then as it is not their breeding season.)

5   *Paramecium aurelia* may be a more efficient feeder. *Lemna minor*, which floats, reduces the light reaching the submerged *L. trisulca*.

Succession and sampling (page 30)

1   Particles provide gaps into which spores and seeds can slip; allow rhizoids and roots to penetrate; give air spaces; hold water.

3   When there are too many organisms to count and when they are distributed relatively evenly in uniform conditions.

4   **i**   Random quadrats and identification; **ii** random line transects/belt transects at a right angle to the river and identification.

Conservation (page 32)

2   High yield involves growth, needing amino acids and proteins. Amino acids contain nitrogen.

3   Fertiliser should not be applied to water-logged soil, nor when it is raining or about to rain.

Meiosis, genetics and gene control (page 36)

1   Meiosis I separates homologous chromosomes; meiosis II separates sister chromatids.

2   Both separate sister chromatids and produce two daughter cells.

3   Portions of non-sister chromatids exchange places, giving new combinations of alleles.

4 2^n combinations. $n = 4$.
$2^4 = 16$ genetically different spermatozoa.

5

	mitosis	meiosis
number of division cycles	1	2
number of daughter cells	2	4
number of chromosomes per nucleus in daughter cells	16	8

Genetics (page 38)

1

parental phenotypes	red flowers	pink flowers
parental genotypes	C^RC^R	C^RC^W
genotypes of gametes	all C^R	C^R or C^W

genotypes and phenotypes of offspring

	gametes from red parent C^R
gametes from pink parent C^R	C^RC^R red flowers
C^W	C^RC^W pink flowers

Equal numbers of red and pink flowers are expected, i.e. a ratio of 1 red : 1 pink.

Dominance and sex linkage (page 40)

1

parental phenotypes	normal ears	short ears
parental genotypes	BB	bb
genotype of gametes	B	b
genotype of offspring (F_1)	all Bb	
phenotype of offspring (F_1)	all with normal sized ears	

$F_1 \times F_1$

genotypes of F_1	Bb	Bb
genotypes of gametes	B or b	B or b

genotypes and phenotypes of F_2

gametes	B	b
B	BB normal ears	Bb normal ears
b	Bb normal ears	bb small ears

A ratio of 3 mice with normal ears : 1 mouse with small ears is expected.

2

parental phenotypes	haemophiliac man	carrier woman
parental genotypes	X^hY	X^HX^h
genotypes of gametes	X^h or Y	X^H or X^h

genotypes and phenotypes of offspring

gametes	X^H	X^h
X^h	X^HX^h carrier female	X^hX^h haemophiliac female
Y	X^HY normal male	X^hY haemophiliac male

A ratio of phenotypes 1 : 1 : 1 : 1 is expected.

Multiple alleles and dihybrid crosses (page 42)

1 Blood groups A and B.

2

genotypes	phenotypes
C^AC^A	agouti
C^AC^g	agouti
C^AC^h	agouti
C^AC^a	agouti
C^gC^g	chinchilla
C^gC^h	chinchilla
C^gC^a	chinchilla
C^hC^h	himalayan
C^hC^a	himalayan
C^aC^a	albino

3 $C^AC^h \times C^AC^h$ or $C^AC^h \times C^AC^a$

4

parental phenotypes	purple stems jagged leaves	green stems smooth leaves
parental genotypes	AABB	aabb
genotypes of gametes	AB	ab
genotypes of offspring (F_1)	AaBb	
phenotypes of offspring (F_1)	purple stems jagged leaves	

$F_1 \times F_1$

genotypes of gametes AB Ab aB ab × same

gametes	AB	Ab	aB	ab
AB	AABB purple s. jagged l.	AABb purple s. jagged l.	AaBB purple s. jagged l.	AaBb purple s. jagged l.
Ab	AABb purple s. jagged l.	AAbb purple s. smooth l	AaBb purple s. jagged l.	Aabb purple s. smooth l
aB	AaBB purple s. jagged l.	AaBb purple s. jagged l.	aaBB green s. jagged l.	aaBb green s. jagged l.
ab	AaBb purple s. jagged l.	Aabb purple s. smooth l	aaBb green s. jagged l.	aabb green s. smooth l

A ratio of phenotypes of 9 purple jagged : 3 purple smooth : 3 green jagged : 1 green smooth is expected.

Test crosses and χ^2 test (page 44)

1 It is a test of heterozygosity.

2

parental phenotypes	coloured fur	albino
	normal sized ears	small ears
parental genotypes	AABB	aabb
genotypes of gametes	all (AB)	all (ab)
offspring genotypes	all AaBb with coloured fur and normal sized ears	
parental phenotypes	coloured fur	albino
	normal sized ears	small ears
parental genotypes	AaBb	aabb
genotypes of gametes	(AB)(Ab)(aB)(ab)	(ab)

genotypes and phenotypes of offspring

gametes	**ab**	
AB	AaBb	coloured fur normal ears
Ab	Aabb	coloured fur small ears
aB	aaBb	albino normal ears
ab	aabb	albino small ears

A ratio of 1 : 1 : 1 : 1 is expected.

3 i Monohybrid test cross: Aa × aa. **ii** Dihybrid test cross: AaBb × aabb.

4 $\chi^2 = 2.67$. $\eta = (2 - 1) = 1$. The probability of $\chi^2 = 2.67$ is between 0.05 and 0.1. The results are due to chance.

5 $\chi^2 = 25$. $\eta = (4 - 1) = 3$. The probability of $\chi^2 = 25$ is < 0.001. The results are not due to chance. The expectation of a 1:1:1:1 ratio is incorrect. Another explanation is needed.

Mutation (page 46)

1 Addition of one nucleotide alters all subsequent triplets. Some may now code for different amino acids but some may be altered to stop signals.

2 The altered triplet may still code for the same amino acid.

5 Those in low light intensity will be less vigorous, less green, have thinner but longer stems.

Gene activity (page 48)

1 Transcription is production of mRNA from DNA. RNA polymerase is needed to join RNA nucleotides into mRNA.

Classification, selection and evolution (page 52)

5 Viruses do not have cells or metabolism.

Allele frequencies (page 54)

3

| *parental genotypes* | H^NH^S | H^NH^S |
| *genotypes of gametes* | (H^N) (H^S) | (H^N) (H^S) |

gametes	H^N	H^S
H^N	H^NH^N	H^NH^S
H^S	H^NH^S	H^SH^S

A ratio of 1 normal homozygote : 2 heterozygotes : 1 sickle homozygote is expected.

4 The mating gives an expected ratio of 1 homozygous for resistance : 2 heterozygotes : 1 homozygous for susceptibility. The heterozygotes (with both alleles) have the best chance of surviving.

5 The *carbonaria* allele is dominant. Moths carrying it are melanic and therefore camouflaged against bird predators in polluted areas, but conspicuous in rural areas. Selection is *for* melanism in polluted areas and *against* it in rural areas.

Speciation (page 56)

1 Isolation allows genetic drift in a small population and selection for different adaptations from those of the main population. When enough differences have built up the population may be reproductively isolated.

2 'Good' species (biospecies) do not interbreed.

3 All the wren populations can still interbreed successfully and so are one species.

Artificial selection (page 58)

3 Increased yield; compact growth; tolerance of poor soils, drought or flooding; resistance to pests and diseases.

4 It is difficult to select for desirable alleles of many genes when choosing individual organisms as parents. There is often a large environmental effect on the phenotype.

Kidney (page 63)

5 Renal artery; afferent arteriole; glomerulus; efferent arteriole; capillaries in cortex and medulla; venule, renal vein.

Nerve cells (page 68)

1 Both are long cells; have centrons and axons; carry nerve impulses; may be myelinated. A sensory neurone has a long dendron; a motor neurone, dendrites. A sensory neurone carries impulses from a receptor to the central nervous system; a motor neurone carries them from the central nervous system to an effector.

Nerve impulses (page 70)

3 Depolarisation makes the p.d. across the membrane *less* negative (and eventually positive) inside with respect to outside. Hyperpolarisation makes the p.d. across the membrane *more* negative inside with respect to outside.

Regulation of plasma glucose (page 76)

1

receptors	effectors
α and β cells; receptors in plasma membranes of liver, muscle and fat cells	liver, muscle and fat cells

Answers to exam-style questions

In these answers a semi-colon (;) separates individual marking points. Notice that in many cases there are more marking points available than there are marks for the question. This means that there are several ways in which you can gain full marks for a question. In the mark scheme, the following symbols may be used: A, accept; AVP, alternative valid point; AW, alternative wording; R, reject.

Energy and Respiration

1 a **i** prevent enzymes/membrane proteins from being denatured; [1]

ii keep acidity/pH constant; enzymes have an optimum pH (that is often near pH 6.8); [1 max.]

iii decreases the water potential/solute potential; prevents water entering organelles by osmosis; to make the solution isotonic with the organelles; [1 max.]

b **i** A, nuclei; B, mitochondria; C, ribosomes; [3]

ii glycolysis takes place in the cytosol/cytoplasmic matrix; does not occur inside organelles; involves soluble cytoplasmic enzymes; [1 max.]

c **i** increase/double the rate of *aerobic* respiration/oxidative phosphorylation/Krebs cycle; because enzyme-controlled reactions proceed faster at higher temperatures; kinetic energy of reacting molecules is greater; [2 max.]

ii no effect; mitochondria do not use glucose (as a respiratory substrate); [2]

iii increase the rate of *aerobic* respiration/oxidative phosphorylation/Krebs cycle; the link reaction takes place on/in the mitochondria; glycolysis normally supplies the mitochondria with pyruvate; [2 max.]

iv increase the rate of respiration/oxidative phosphorylation/Krebs cycle; if more ADP is available it can be used to make more ATP; [2]

v increase the rate of respiration/oxidative phosphorylation/Krebs cycle; NAD is reduced during the Krebs cycle; [2]

2 a from pyruvate; which is decarboxylated to form ethanal; allowing reduced NAD to be converted to NAD/reoxidised/recycled; which allows glycolysis to continue; ref. to anaerobic respiration/fermentation; [3]

b **i** small molecule; very soluble in water; soluble in lipids/can pass through phospholipid bilayer; rapid diffusion; no need for a carrier protein; AVP; [2]

ii electron transport chain; in the mitochondria; or by any enzyme controlled reaction that requires hydrogen; in the synthesis of fatty acids; [2]

3 a **i** RQ is a ratio; it gives the relative rate at which oxygen is used and carbon dioxide is produced; [1]

ii $0.69 = 12.0 \text{ mm}^3 \text{ min}^{-1} \div$ rate of oxygen production; rate of oxygen production = $12.0 \div 0.69 = 17.4 \text{ mm}^3 \text{ min}^{-1}$; [2]

iii species C and D; because both have an RQ close to 0.7; [2]

b all variables must be constant if the species are to be compared fairly; gases/air expand and contract with temperature changes; the respirometer would therefore give false readings; enzyme activity alters with temperature; so the rate of respiration would change; the solubility of oxygen in water decreases with increasing temperature; AVP; [4]

Photosynthesis

1 a oxygen was produced during photosynthesis; increasing the pressure in the syringe/apparatus; [2]

b movement in opposite direction/towards the syringe; because the cells would absorb oxygen during respiration and the carbon dioxide produced would dissolve in the water;

or no movement of the bead; because the volume of oxygen used in respiration would be balanced by the carbon dioxide produced; [2 max.]

c **i** the rate of movement would increase steadily/linearly as the temperature increased; because when the cells are in bright conditions, the Calvin cycle/carbon fixation/light-independent reactions limit photosynthesis; these enzyme-controlled processes go faster at higher temperatures; 20°C is unlikely to cause damage/denaturation; AVP;

ii rate would only increase to a small/smaller extent; light dependent reactions/synthesis of ATP/reduced NADP would be limiting; bead might start moving towards the syringe at higher temperatures; because respiration would increase/become faster than photosynthesis; AVP; [5 max.]

d **i** ^{14}C incorporated into organic compounds/combines with RuBP; converted to other substances/sugars in the Calvin cycle; chain of enzyme controlled reactions; [2 max.]

ii glycerate phosphate (PG); [1]

2 a more palisade cells can be packed closely under the upper epidermis; nearer the light; substomatal air spaces do not interrupt the palisade mesophyll; stomata on the upper surface more likely to be blocked by dust/rainwater; the atmosphere will be more humid under the leaves; so less water will be lost in transpiration from stomata on the lower surface; AVP; [3]

b stomata can shut; the cuticle is not very permeable to water/waxy; stomata may be sunken; hidden under a layer of hairs; so that the diffusion gradient is reduced; the leaf may be shiny/reflective; so it does not become hot; AVP; [2]

c **i** chloroplasts can move to places in the cell with optimum light intensity;

ii maximum chance for light/photons to be absorbed (by a chloroplast); large surface/volume ratio;

iii light passes down cell for a maximum distance; [3]

d temperature; humidity; water supplied to cuttings; day length; concentration of nutrients/ions/named nutrient; [2]

e the genotype is constant/no genetic variation; all cells produced by mitosis; the plants/leaves are from one clone; [1]

f **i** are below the light compensation point; respiration is faster than photosynthesis; [2]

ii in the cells developing at low light intensity (or converse argument), lower concentrations of enzymes; fewer/lower density of mitochondria; fewer/lower density of chloroplasts; less chlorophyll; lower concentrations of NAD/NADP/P_i/ADP; reduced rate of respiration; becomes light saturated at low light intensity; at intensities above about 800 lux; [4]

3 a **i** products of the light-dependent reactions; photophosphorylation; Z scheme; ref. to photosystems I and II; in grana (of chloroplasts); [3]

ii used again in the light-dependent reactions/recycled; [1]

b RuBP is constantly used and reformed/a stage in the Calvin cycle; [1]

c triose phosphate is transported out of the chloroplast, condenses to form hexose/glucose phosphate; from which glucose, starch (and sucrose) are produced; [2]

Populations and Interactions

1 a combine with cell membranes; combine with/denature proteins/enzymes; [1 max.]

b woodlice – 3:1; springtails – 1:3; [2]

c **i** the larger number of litter animals/detritivores in Wetmoor Wood break down the litter more quickly (AW); bacteria/micro-organisms/fungi may be inhibited in Hallen Wood; [1 max.]

ii the cadmium/heavy metal has adversely affected a parasite/predator of springtails; springtails are tolerant/resistant (R immune) to cadmium; competition between springtails and other detritivores is less; [2 max.]

d **i** plants 5.0 + 3.8 + 14.2 = 23.0 ÷ 3 = 7.67 mg kg^{-1}; animals 56.5 + 119.3 + 125.7 + 41.1+ 56.6 = 399.2 ÷ 5 = 79.8 mg kg^{-1}; [2]

ii animals eat more than their own mass/weight of plant; reference to pyramid of biomass; animals at higher trophic level concentrate materials gained from lower levels; [2 max.]

iii different niche; feed on different things; active/develop at different times; occupy different microhabitats; live for different lengths of time, longer life gives more time to accumulate cadmium; may differ in ability to absorb/excrete cadmium; AVP; [2 max.]

e reference to quadrats/point quadrats; grid and use of random number table; suitable measure of abundance (frequency/density/% cover); AVP; [2 max.]

f there was originally a low concentration of cadmium in the wood; concentration increased (as a result of pollution from smelter); original population sensitive to cadmium; very low frequency of genes/alleles/genotypes that are tolerant; reference to mutation giving rise to tolerant genotypes/alleles conferring tolerance; mutation random/ not caused by the cadmium; tolerant genotypes have no advantage/may be disadvantageous when cadmium concentration is low; tolerant genotypes survive/grow relatively well in presence of cadmium; therefore become more frequent; reference to selection/natural selection; recombination/ crossing between different tolerant genotypes, may result in greater tolerance; as cadmium concentration increases over many years, alleles producing tolerance completely replace those that result in sensitivity; AVP; [6 max. + 1]

Meiosis, Genetics and Gene Control

1 a long-rayed plants breed true/produce only long-rayed offspring; no-rayed plants also breed true/produce only no-rayed offspring; short-rayed plants produce offspring of all three types; in the ratio of 1 long : 2 short : 1 no ray; [3 max.]

b suitable genetic symbols, e.g. R^n = no ray and R^l = long ray; long ray and no ray plants homozygous; both these produce only one genotype of gamete/eggs/ovules/pollen; short ray plants are heterozygous; produce two genotypes of gamete; codominance of the two alleles; genetic diagrams; [5 max.]

c all short ray; [1]

d cross-pollination/fertilisation of one long-ray ovule/egg by a single no-ray pollen grain; a short-ray seed getting into the wrong container; mutation (of a long-ray allele to a no-ray allele); AVP; [2 max.]

2 a 1 round : 1 wrinkled [1]

b for round grains, 25.5; for wrinkled grains, 7.3; [2]

c move slide without looking down microscope, measure grain nearest the graticule division/pointer/centre of field; [1]

d stain with iodine solution; [1]

e different cells receive different amounts of sugar/carbohydrate; environment in which the grains form is variable; not all cells produce the same amount of enzyme; grains may not all be same age/stage of development; AVP; [2 max.]

f alpha; [1]

g boil the extract and then show that it will no longer produce starch; change the pH and demonstrate that the reaction has an optimum pH; [1 max.]

h ATP; [1]

i concentration/amount of glucose-1-phosphate; volume of extract used; mass of peas used to extract the enzyme; temperature; pH; AVP; [2 max.]

j structure of enzymes depends on genetic code/genes/genotype; reference to the nucleotide sequence of the DNA; round and wrinkled seeds have different alleles; therefore code for slightly different proteins; which synthesise starch at different rates; round allele codes for more active/efficient enzyme; [2 max.]

Classification, Selection and Evolution

1 a dogs and wolves are in same genus/chimps and humans in different genera; because dogs and wolves are thought to be more similar/more closely related; [2]

b the criterion of a species being a group capable of interbreeding cannot be applied; so less clear-cut evidence of structural and biochemical similarity must be used; difficult to distinguish between varieties and species; [2]

c tradition; to understand evolutionary history; help identify specimens/allow researchers to be confident that they are referring to the same type of organism (when discussing/reporting investigations); allow logical and consistent names to be given; help understand organisms by comparing and contrasting them with others; to encourage conservation; AVP; [3]

2 a environment on Orkney and on the mainland different; colder/windier/more exposed/less fertile/other examples of difference; fewer/different predators on islands; larger voles more vulnerable to predators/less vulnerable to cold; differences between the populations arose by mutation; by recombination of genes; variants/phenotypes best fitted to survive in each environment left more offspring; selection in opposite directions/disruptive selection; Orkney voles descended from very few original colonists/founders; small populations allow genetic differences to arise by chance; reference to gene/genetic drift; little/no chance of new (vole) colonists from other places reaching Orkney; so populations are reproductively isolated; no interchange of genes/merging of genetic differences; comparison with other island groups, e.g. Galapagos; AVP; [7+1]

b a species is a group of very similar organisms; can interbreed freely; share a common gene pool/exchange genes; Orkney voles similar to mainland ones, so should be regarded as same species; but have set of distinctive characteristics/phenotypes, so can be regarded as different species; unknown if they could produce fully fertile offspring if crossed; no opportunity for selection to favour individuals that failed to cross/prefer to mate within their own population; AVP; [5]

3 a during meiosis; reference to prophase I/first division; during the production of sperm cells/male gametes; [2]

b they could be close relatives (with many genes in common); e.g. brothers always have their father's Y chromosome/similar example; mutation is rare; so there could be many generations separating them from their common male ancestor; ref. to independent assortment (of the other chromosomes/autosomes); AVP; [3]

c humans may all have descended from a small number of women/small ancestral group/small population; chimpanzees are a more ancient species; giving more time for diversity to develop; natural selection could have favoured greater diversity in chimpanzees; [2]

d increases the amount of variation; by producing new combinations of alleles; a (new) mutant allele may be unfavourable in one genetic background but favourable in combination with another genotype; genes interact with other genes to produce a phenotype; [3]

e there is no recombination/crossing over; so the nucleotide sequence remains unchanged (for many generations); only mutation can change it; [2]

f showing a fundamental similarity; the same nucleotide/genetic sequence; having a common origin; able to pair during meiosis; reference to an example of anatomical homology such as the pentadactyl limb; [2]

4 artificial selection occurred; easy detachment/brittleness allows the seeds of wild grasses to be shaken loose/dispersed; by wind/animals; humans must gather the seed of domesticated varieties/cereals; strongly attached/non-brittle ears will be successfully gathered without loss of seed; mutations/variants in which the ear was (more) strongly attached had an advantage (when the seeds were harvested and artificially sown); mutant genes/alleles producing strongly attached ears spread in domesticated populations; are eliminated from wild populations; AVP; [5]

Control, Coordination and Homeostasis

1 a the radioactivity is increased; by about 1000 counts min^{-1}; [2]

b axon/axon membrane became more permeable to sodium ions; sodium ions entered/rushed into the axon; because of a concentration/electrochemical gradient; reference to an axon potential; [3 max.]

c second–third stimulus, radioactivity declined/was reduced from 1400 to 600 counts min^{-1}; fifth–sixth stimulus, radioactivity declined/was reduced only slightly/not significantly; [2]

d active transport is prevented; because active transport requires ATP; sodium pump/sodium–potassium pump does not function (in presence of inhibitor); action potential/influx of sodium ions does not require ATP; response of the axon to stimulation not affected by the inhibitor; [4 max.]

2 a hormonal slow acting, nervous fast; hormonal long lasting, nervous produces short-term/behavioural effect; hormonal influences many cells, nervous may influence a single cell/group of cells; nervous involves action potentials/electrical impulses, hormonal involves a chemical substance carried by blood; AVP; [3 max.]

b seeds soaked in water, 10 mm^2; seeds soaked in the presence of embryos, 16 mm^2; [2]

c seeds produce amylase; that hydrolyses/digests the starch; producing sugar/maltose; storage tissue/endosperm only produces a small amount of amylase in absence of the embryo; embryos produce a hormone/growth substance that stimulates seeds to produce amylase; reference to gibberellin; AVP; [5 max.]

Index

Note: page numbers in **bold** indicate major topics.

A-level Study Guide

Media Studies

Jacquie Bennett

Revision Express

Acknowledgements

Film ratings icons on p. 125 reproduced by permission of the British Board for Film Classification. The BBFC classification symbols are both copyright and trademark protected.

Screenshot (p. 33) reprinted by permission from Microsoft Corporation.

Photograph on p. 41 taken by Kerry Vines.

Series Consultants: Geoff Black and Stuart Wall

Pearson Education Limited

Edinburgh Gate, Harlow

Essex CM20 2JE, England

and Associated Companies throughout the world

www.pearsoned.co.uk

© Pearson Education Limited 2005

British Library Cataloguing-in-Publication Data

A catalogue entry for this title is available from the British Library.

ISBN -10: 1-4058-2366-6
ISBN -13: 978-1-4058-2366-1

First published 2005

10 9 8 7 6 5 4 3

09 08 07 06

Set by 35 in Univers, Cheltenham

Printed by Ashford Colour Press, Gosport, Hants

Key concepts

All of your study in Media Studies will be constructed through analysis of the key concepts. You need to be able to deconstruct media texts (media language) quickly and easily and you need to be able to do this in relation to these key concepts.

Key concepts are the main factors which contribute to the construction and deconstruction of a media text. By equipping yourself with the skills to analyse texts in relation to these key concepts, you will be able to apply them to any media texts with which you engage.

In this first section, therefore, we look at what each of these key concepts are and how they can shape a media text. In the next section we will focus on how you can use this knowledge of the key concepts to analyse a range of media texts.

Exam themes

- ➜ Media language
- ➜ Representation
- ➜ Audience
- ➜ Ideology
- ➜ Institution
- ➜ Narrative
- ➜ Genre

Topic checklist

○ AS ● A2	OCR		AQA		WJEC	
Key concepts	○	●	○	●	○	●
Media language	○	●	○	●	○	●
Representation	○	●	○	●	○	●
Audience	○	●	○	●	○	●
Ideology	○	●	○	●	○	●
Institution	○	●	○	●	○	●
Narrative	○	●	○	●	○	●
Genre	○	●	○	●	○	●

Key concepts

In Media Studies we base all the study around a range of key concepts, as you know. Why do we do this?

Media Studies is a very fast moving subject and covers a huge amount of information across a number of different areas of study – you need practical skills; you study bits of sociology, psychology, economics, history, cultural studies, visual studies, graphic design. . . . In order to put all this into a form which you can cope with, all the exam boards focus everything around the same range of key concepts.

What are the key concepts? ●●●

Although there are some differences between the boards, the basic key concepts are common to all specifications at this level:

→ media language
→ audience
→ representation
→ institution
→ ideology.

The other important key concepts which you need to learn about are:

→ narrative
→ genre.

With this focus on these key concepts, you are better able to develop a broad and deep understanding of the connections between prior knowledge; understanding of contemporary media forms; learning about other media forms and understanding of production work, which is the intention behind all these specifications. At A2 level, you will be better able to develop the required knowledge and understanding of a range of media industries and their institutional, technological and global contexts with a 'synoptic' understanding of the media and the ability to engage with critical autonomy with a range of texts if your study is grounded in these guiding principles for study.

Media language

Media language is the starting point for all study of the media and all media production work. Unless you understand how a media text communicates to an audience and how to construct texts that communicate appropriately to an audience you cannot get far with Media Studies!

You need to be able to understand how different techniques are used to create meaning (such as the use of colour or the use of the camera in a moving image text) and how technical features can convey a sense of character, of themes and of emotional context for a text, for example. There are many ways of doing this – in this section we look at the basic ways of 'reading' a media text. In the next section (Reading the media) we look at ways that you can do this in more depth by using specific forms of analysis such as **semiotic analysis**.

Checkpoint 1

What do we mean by a key concept?

The jargon

Critical autonomy means that you are able to discuss and write about texts independently, using your own ideas. You may make use of debates, theories and concepts but in your own words, demonstrating that you have thought about the texts you are using.

Examiner's secrets

Examiners like to see critical autonomy in your answers, rather than learned answers. Often the best answers are very individual and examiners are in no doubt about the candidates' opinion of the texts!

Media audiences

Any media text is created for a particular **audience** and will usually appeal most to this particular target audience. You need to be aware of ways that audiences can be categorised and how the makeup of the target audience affects the **media language** employed by a text.

Media representations

All media texts are **re-presentations** of reality and you will need to be able to identify what representations of people, place, time, etc are being established within a text and how these are employed to create meaning. They may be **stereotypes** or they may be complex representations but it is important to remember that they have all been constructed to appeal to a particular **target audience**.

Media institutions

A media text is made by a particular media institution and this will also affect the way that it is constructed and the meaning it communicates. It may also affect audience expectations. A new release from a well-known rap star called 'Getting down' will immediately establish different expectations in an audience than a new release from an unknown boy band by the same name. The **'institution'** of the rap star making the record has affected the way meaning is constructed, the **representations** employed and the **audience expectations** for the track.

Media ideology

The **'ideology'** of a media text relates to the **values and attitudes** employed within the text and conveyed to the audience. These may be **explicit** (i.e. obvious) or **implicit** (not obvious but underpinning all aspects of construction). It is important to identify the ideologies operating within a media text when you are studying or deconstructing it, so that you can understand **why** it has been constructed in this way and why it is aimed at a particular audience.

Narrative and genre

You need to be able to understand the way a media text has been structured and how the audience is **'hooked'** into the text as well as the **codes and conventions** which are used in the text. These establish the **'type'** of text (e.g. action/adventure film or docusoap) and affect **audience expectations** and how the **ideologies** are communicated to the audience.

Examiner's secrets

Remember that it does not matter if you identify the 'wrong' target audience for a text – it is your ability to justify that choice with evidence from the text which is important.

Checkpoint 2

What is a soap opera?

Exam question   answer: page 20

Choose a text which you know well (such as an episode of your favourite soap opera) and see if you can identify each of the key concepts and how they are operating in the text. Can you see how these key concepts are all interrelated?

(1 hour)

Media language

Media language means the way in which a text is constructed to create meaning for a reader or viewer of the text. In the 'Reading the media' section of this book we look at how you can 'read' the language of a media text. In this section we will look at some of the main ways that you can structure this reading of a text. You can explore how meaning is communicated by or through a media text in many different ways. Some of the most important ways of 'reading' a text are listed here.

Checkpoint 1

Do you think all institutions have the same level of control over the texts they produce?

Links

If you look at the section about audience theory on pp. 86–91 you will see that the relationship between institution and audience is also analysed in terms of active and passive participation there. All key concepts are very closely related in Media Studies, which is why you must always cover them all in your analysis.

The jargon

The *logos* of a text is the wider contexts which surround the text and influence the 'meaning' of the text.

Checkpoint 2

Are reality TV programmes produced by institutions and fed to passive audiences or do active audiences demand these texts?

Text as interface between institution and audience

The text acts as an interface between the institution that has created the text and the audience that will receive the text. There are three basic ways of looking at the role of the text in this relationship.

➜ An **active institutional** view is that meaning is 'transmitted' by the text from the institution to the audience. The institution is seen as **active** and the audience as **passive** receiver.

➜ A **negotiated** view is that meaning is constructed by the text. The institution encodes a meaning into the text and the reader interprets this in relation to other factors as well (for example, knowledge of previous texts). Meaning is therefore '**negotiated**' between institution and audience.

➜ An **active audience** view is that meaning is re-created by the audience – so the institution becomes **passive** in the relationship since it has no control over how the audience re-creates meaning from the text in an **active** way.

Deconstruction

Jacques Derrida explored the relationship between institution and audience in a different way. He suggests that an audience deconstructs a text within the context of its **logos**. Therefore there is not a single meaning to a text but many meanings and many interpretations based on these contexts and the way the audience receives the text – for example, by making use of previous knowledge of related texts or a particular social context for production.

Reading media language

When you are deconstructing a media text – especially during an exam – you are most likely to do so in terms of the sign system being employed to create meaning for the audience. However, when thinking about the concept of media language, it is important to remember that mass media texts are not created or consumed in isolation; therefore you should always consider these contexts and how they affect the construction and consumption of the text.

If you are studying this text from an **active institutional** point of view – i.e. assuming an active institution – you would be first asking questions such as:

→ Who constructed this text?
→ What context did they construct it for?
→ What other texts have they constructed?
→ What codes and conventions can I recognise from other texts they have constructed?

If you are studying the text from a **negotiated** point of view – assuming a negotiated reading – you would be asking questions such as:

→ What genre codes and conventions are being employed here?
→ What do I know about the time and place where this text was constructed?
→ Is this text typical of its genre or time and place?
→ What representations are being used in this text to create meaning?
→ What meaning has been encoded into the text?
→ What codes and conventions are being used in the text?

If you are studying the text from an **active audience** point of view – assuming an active audience – you might be asking questions such as:

→ How does this text conform to audience expectations?
→ What previous experience does the audience use when consuming this text?
→ How does the audience create meaning from this text?
→ How and where might an audience receive this text?
→ How might this influence the meaning they receive?

A detailed reading of a text might well deconstruct the text in all these terms and then consider the text carefully in relation to its **logos** before reflecting on a range of media issues and debates raised by the text. You can see why the key concepts are all interrelated!

Exam question   answer: page 20

Watch a music video for a current Top 10 hit. Give a reading of the text to include active institution, negotiated and active audience analysis, taking care to deconstruct the text in relation to its logos as well.   (1 hour)

Representation

Representation relates to the representation of reality in the media.

When we study representation we look at what representations are constructed, why they are constructed that way and how audiences receive and consume them. In many situations our only knowledge of people or situations will come from the media, so these representations are very important. We are very unlikely to meet Michael Jackson, for example, so our whole knowledge of him comes from the presentations in the media. We can only judge by the representation we are given in the media.

Media representations

It is impossible not to receive these representations from the media but it is not impossible to become aware of how we receive them and to make conscious decisions about why we accept or challenge them. This is why representation is a key concept. Just as with media language, we are giving you the tools to make informed autonomous decisions about the media you consume.

It is also impossible for these representations to be unbiased in any way. Whatever representations are used, there an **ideology**, **meanings** and **values** are implicit in that presentation. Clearly, this gives great power and responsibility to media institutions – for example, the presentation of asylum seekers in some newspapers over the last few years has been very negative and this can be seen to have influenced social attitudes in many ways. Newspapers, in particular, are often accused of starting **moral panics** by the way they represent people or groups.

Representation of social groups

We often analyse representations in the media according to categories such as:

→ age
→ disability
→ gender
→ socio-economic grouping
→ race
→ nationality
→ sexuality.

We tend to analyse how 'older' people or 'Native Americans' are portrayed or how 'women' are represented, for example.

Stereotypes

Stereotypes are characters in a media text who are 'types' rather than complex people. Stereotypes are usually defined by their role – (e.g. bad cop or family matriarch) and will not usually be developed in any depth. Children's media texts often make extensive use of stereotypes because they enable a young audience to identify very quickly with the action and situation. *Brum* is a good example of a programme which uses a central non-human character and a range of stereotypes to create meaning.

Checkpoint 1

Ideology and representation – If you were constructing a new documentary about school leaving age, what representations would you use if your programme wanted to suggest that it should be lowered to 12?

Watch out!

There are many different ways of 'categorising' people and places in terms of representations – this is not a complete list!

Examiner's secrets

If you are writing about the use of stereotypes in a text, make sure you do write about their use, not just describe how they are constructed!

Boy bands are often accused of being constructed to offer a calculated range of representations, each to appeal to different elements of the audience, e.g. the sporty one, the casual one or the quiet one.

Stereotypes are usually negative representations and most have a lot of assumptions and 'cultural capital' invested in them – again usually for negative effect. Stereotypes also usually represent an entire social group in a single character – such as the naïve secretary, the spoilt child or the drunken Irishman.

Ideology

As you can see, the representations used within a text act to define the ideology. A family sitcom depends on the use of stereotypes to find the comedy, for example, and each family member's character and role is defined before the programme even starts! Because we know the ideology of the programme, we know what representations to expect – that's why a comedy programme which successfully subverts these can be even more successful – such as *Absolutely Fabulous*!

Genre

Representations are almost always defined by the genre. You would not expect to see an unattractive heroine in a Hollywood action/adventure movie, or a policeman who convicts the wrong person in a police drama. Equally, there are particular stereotypes which we associate with certain genres – the 'matriarch' in a soap opera, the policeman with drink problems and family problems in a crime series, for example.

Audience

The relationship between audience and representation is complex. Do the representations used in a particular text attract a certain audience and thus create a target audience for the text? Or do the audiences expect particular representations in a text? Certain audiences can have a great deal of influence on representations used – a good example is the representations constructed in recent reality game shows where these representations (often bordering on stereotypes) empower an audience to make decisions about who to 'vote out'.

Institution

All representations must also be analysed in the context of the institution creating the representation, e.g. the BBC, as a PSB organisation, has very strict controls about representations of disadvantaged social groups and has a remit to offer 'positive representations' where possible.

Types of representation

Of course representation as a key concept is far wider than simply the representation of the groups highlighted above. You also need to think about representation of different religious groups (think about Muslim representations in the UK news post September 11 and the Iraq war), different cultures and societies, nations, occupations and societies. Above all else, you need to think about **why** a group is being represented in this way in this text.

Links

You will see that each of the key concepts is interrelated with the other key concepts. This is why we don't examine each one in turn but expect you to use them all in all your answers!

Checkpoint 2

Think about a genre which you are familiar with and identify four stereotypical roles associated with that genre. In what ways might these be said to 'define' the genre?

Links

There is a lot more about audience later in this book (see pp. 86–91) and that will help you to think about this complex relationship.

The jargon

PSB means 'Public Service Broadcaster'. This means that the BBC is funded by the licence fee all TV owners must buy. Because of this, the BBC must provide a particular range of programmes to appeal to a very wide audience and must not be seen to be biased in its representations.

Exam question   answer: pages 20–21

Watch an evening news bulletin. What social groups are represented in this bulletin? In what ways are these representations used to create meaning for the audience?   (1 hour)

Audience

For many people, audience is the most important of all the key concepts. If you think of media texts as being like paintings or statues, which are produced for a purely artistic motive, without any thought of an audience, then there is no need to consider audience as a key concept. However, **mass media** texts differ greatly from most fine art since there is always a commercial aspect to production and thus the text is made within a particular context. Part of that context has to be the structuring of a text to appeal to a particular target audience. To use the art example again, it is like the artist who is commissioned to produce a painting of a child by a parent. If the child is not **represented** in a positive way, the client will be unhappy and the painter will not be paid. Knowledge of the target audience (i.e. the parent) defines the ideology of the painting which the painter will have to use (happy, attractive child) and thus the representation to be used.

The consumer

We are all consumers of media texts every day and it is our consumption which is of interest to media producers. They need to ensure they can readily define a target audience for a particular product and then create a product which will appeal directly to that target audience. The success of a media text is directly judged in terms of audience size, so media producers have a good reason for wanting to attract large audiences. This may be obvious for a newspaper proprietor, for example, who can judge success by the number of copies of the newspaper sold, but less obvious for broadcast media. However, a successful commercial television programme will have a larger audience and thus attract more advertising and a successful PSB text with a large audience is justification for the continuation of the licence fee system.

Types of audience

Mass audiences
Mass audiences are, basically, large audiences – for example, the audience for *Eastenders* or for a Premiership football match. Mass audiences are often termed 'broadcast' audiences, who consume mainstream or popular culture texts such as soap operas, sit-coms or reality TV shows.

Niche audiences
A niche audience will be much smaller than a mass audience but usually very influential. Niche audiences are often very dedicated and loyal and thus may still be very attractive to advertisers (especially of relevant products) or sufficiently reliable to enable a niche publication to continue due to the consistent revenue being generated by sales. For example, the introduction of BBC Four is aimed at a niche audience who are interested in artistic programmes.

Checkpoint 1

What media texts do you consume every day? Do you have certain types of text that always appeal? How do the media producers target you?

Checkpoint 2

Look at an edition of a local weekly newspaper. In what ways does it carefully target a niche audience?

Niche audiences consume narrowcast items such as farming programmes or American football.

Audience demographics ●●●

One of the most common ways of identifying a target audience is the socio-economic model. Although this model has been in use for a very long time, it is still a useful tool in identifying a **target audience** and thus **deconstructing** a text. The basis for this system is money – an AB audience, for example, is assumed to have more spending power than a CDE audience. It is also presumed to consume different media texts, such as 'high culture' texts, e.g. broadsheet newspapers and late-night art programmes on TV!

➜ Group A – lawyers, doctors, scientists, managers of large-scale organisations – well paid professionals.
➜ Group B – teachers, senior managers, some middle management – fairly well paid professionals (and very poorly paid teachers!).
➜ Group C1 – 'white collar', junior management, bank clerks, nurses.
➜ Group C2 – skilled 'blue collar' workers such as electricians, plumbers, carpenters.
➜ Group D – semi and unskilled manual such as drivers, post sorters.
➜ Group E – students, the unemployed, pensioners.

Audience profiling ●●●

There are many other ways that an audience can be segmented or profiled. Whenever a media producer is creating a text, they will always take care to identify the target audience as accurately as possible. After all, if the audience is identified and targeted well, the product is likely to be successful! That is why audience is considered a key concept – because it is fundamental to the planning and shaping of a media text.

Audiences can be **segmented** in other ways as well as the socio-economic model:

➜ age
➜ gender
➜ demographic (i.e. where the audience live)
➜ profiling (this is often done by advertisers to identify 'types' of consumers)
➜ values, attitudes and lifestyles (i.e. segmenting audiences by 'tribe' – in other words the types of texts which are likely to appeal to those with these values, attitudes and lifestyles).

Identifying the target audience

So, as you can see, it is very important to identify the target audience for a text you are studying because this target audience will have had a major effect on the shaping of the text and the form and content.

The jargon

High culture is usually defined as intellectual texts such as broadsheet newspapers, programmes on BBC Four or opera/ballet performances. *Popular culture* (the opposite) usually includes tabloid newspapers, gameshows and soap operas, for example.

The jargon

Segmenting an audience simply means dividing it up into different bits.

Examiner's secrets

A lot of students only identify an audience very broadly when writing about texts, e.g. 'it appeals to women' without stopping to think that this might be a generalisation (all women??). Better students can always identify the target audience in much more detail.

Exam question   answer: page 21

'Using a listings magazine or a newspaper, look at the scheduling for two terrestrial channels and two cable/satellite channels. In what ways are you able to identify the differing target audiences?   (1 hour)

Ideology

Ideology is a very important key concept because it underpins the construction of any media text. A media text will always have an ideology which is communicated explicitly or implicitly to an audience. A newspaper will usually have an explicit political ideology, for example, but may have an implicit ideology as well. A film may not appear to have a particular ideology but may be constructing a very strong ideology for the audience. For instance, *Dances with Wolves* was created by Kevin Costner to represent particular ideologies of Native American history within the construct of a conventional Western.

The jargon

A *dominant ideology* is one which is accepted and understood by the majority of people as part of our culture and expectations.

What is ideology? ●●●

In Media Studies the concept of 'ideology' refers to a system of belief that is constructed and presented by a media text. The overarching ideologies that underpin this are the dominant ideologies of our society. The dominant ideologies underpinning a soap opera constructed in the UK are going to be different in many ways from those underpinning a US soap opera and may be very different from, for example, a Chinese soap opera. The media can construct our views of our society and our attitudes towards society.

Think of a news story about a strike. The choice of discourse used in the report about workers 'demanding' a pay rise or management 'conceding' to demands implies an underlying ideology about who is right.

Checkpoint 1

Watch a news bulletin and make careful notes of the discourse used in the reports. How does the reporting establish who is 'right' or 'wrong' in the reports?

Hegemony and dominant ideologies ●●●

Gramsci defined hegemony as the way in which those in power maintain their control. Dominant ideologies are considered **hegemonic** – i.e. power in society is maintained by constructing appropriate ideologies which are usually promoted via the mass media. A Marxist view would be that we are indoctrinated by the media into the views and attitudes that the ruling classes want us to accept.

An accepted **hegemony** in our society is that the police are always right. However, this does not mean that the media indoctrinate us – after all, we often hear news stories where the police are criticised. The media are used to persuade us to accept these dominant ideologies, not to force us to accept them.

Patriotism is a key **dominant ideology** which is promoted through the media – think of the ways the media reports a royal wedding, or the focus on international football or rugby matches.

The jargon

Hegemony is the control of the current society by government through cultural and social expectations as well as the media.

Althusser argued that ideology is a force in its own right. Class rule is sustained by organised power, which operates in two ways at the same time. The military, police and legal system maintain state force by power. Education, religion, politics, society and the media maintain power at the ideological level, by reinforcing 'common sense' assumptions, attitudes and expectations that are ideological in that they manufacture and maintain consent to the existing social order.

Chomsky argues that the mass media can be used to divert people's attention from real issues such as poor living conditions. He maintains that most people prefer to escape into popular culture (such as soap operas or reality TV shows) rather than watch hard-hitting documentaries, for example.

Institutional ideologies ●●●

Of course, there are other ideologies at work in a media text. Any media institution will have particular ideologies which affect the construction of its texts. These may be **explicit** ideologies (e.g. a newspaper may have a particular political affiliation) or **implicit** ideologies (such as the visual representations of a heroine in a typical Hollywood film).

Identifying ideology ●●●

Ideologies are promoted in many ways. Think of the ways that women are portrayed in films, on television and in magazines. Magazines are a very good example because so many present an 'ideal' lifestyle for their readers to aspire to. This 'ideal' is directly linked to the dominant ideologies in society.

When you are analysing a media text it is important to consider that text in relation to the ideologies underpinning the text.

→ Are they explicit or implicit?
→ How does the text employ dominant ideologies?
→ What ideological assumptions are made in the text?
→ Are these ideologies cultured, institutional or both?
→ Do the ideologies reflect the type of text? (For example, the ideologies being established in a washing powder advert may well shape the form and content of the whole advert!)

Checkpoint 2

In what ways can the media build up patriotic ideologies for an international sporting event?

Examiner's secrets

When you are writing about ideologies in a text, don't feel that you need to explain terms like 'hegemony' – your examiner is likely to know what this is!

Examiner's secrets

It's a good idea to think about the ideologies operating in any texts which you are studying in an exam and how these are communicated to the audience. Being able to write about ideologies often gets you extra marks!

Exam question   answer: pages 21–22

Look at the front covers of two lifestyle magazines aimed at different niche markets. To what extent does each cover promote dominant ideologies in our society? How does the magazine make these into aspirations for the reader?

(1 hour)

(You might want to compare a mainstream magazine and one which subverts the mainstream to some extent, such as *Gay Times*.)

Institution

In one sense it seems obvious that institution is a key concept. After all, a media text has to be made by somebody, so there is likely to be an institution involved in the process. Study of institutions is very important, however, because this role also shapes the form of any text being made. In order to analyse the relationship between the text and the target audience, we need to think about how this influences both the production and the receiving of the text. You need to be able to identify the institution which has made the text (for most unseen exercises you will be told the name) and to know enough about that institution to be able to assess how this has affected the way the text is shaped.

What is an institution?

Put simply, the institution responsible for a text is the body which has made the text. You read earlier that the main difference between a mass media text and a fine art text is that the mass media text is produced with a particular context and audience in mind, whereas a fine art text is produced by an artist with no concern about audience or context for consumption. While this is obviously very simplistic, it is a good place to start thinking about the role of the institution. Some media institutions are more heavily commercial, while others have a requirement to produce particular types of text, and to support minority audiences, for example. By being aware of how institutional perspectives affect the **shaping** of a text in this way, you can become better aware of the text itself and the relationship to the target audience.

The shape of a media institution and the texts it creates is usually formed by a trade-off between four crucial influences:

→ money (or lack of it)
→ ownership (and the degree of control this involves)
→ artistic or social motivations (wanting to produce certain texts)
→ target audience (needing to attract a particular audience or being dependent on sustaining a particular audience).

For example, a broadcast institution may be required to provide a certain number of factual programmes or programmes for a particular social group. This would have a significant impact on scheduling and on the production process. A small independent producer may want to make big budget action/adventure movies but may have limited funds and thus have to compromise on special effects.

When you are thinking about institution as a key concept it is important to remember that you need to be able to identify:

→ what institution created a particular text
→ how this **institutional context** has shaped the text
→ what **ideologies** are therefore established in the text
→ what **institutional codes and conventions** are used in the text.

Checkpoint 1

The *Blair Witch Project* is famous for being an 'independent' text which hit the big time. In what ways did their marketing strategy help this?

The jargon

Institutional context is the ways in which a particular text is constructed within a framework defined by the institution which created it.

The jargon

Institutional codes and conventions are the influences of the institution that creates a particular text and will be revealed by certain symbolic codes that you can deconstruct when looking at the text. Some of these will be conventions, i.e. typical codes for this particular institution, such as a jingle played during a radio programme.

Types of institution ●●●

Media institutions can broadly be divided into three types:

→ commercial
→ public service
→ independent.

Commercial institutions

Commercial institutions have to make money to survive – newspapers have to sell advertising space and sell copies; broadcasters have to generate advertising revenue. So, they cannot afford to produce texts which will attract small audiences – especially niche audiences – as the advertisers will want access to large and stable audiences. This is why most commercial institutions will target a 'mass' audience rather than risk losing the advertisers. Thus ITV3 shows programmes calculated to bring in big audiences, such as *The Bill*.

Soap operas, reality TV shows and quiz shows are especially popular for commercial television providers because they focus on ordinary people and so help the audience to engage with the programmes by identifying with the participants. This might be another reason why other institutions, such as the BBC, then catch onto similar programmes, to try and attract back the same audience – which is why critics can then accuse the BBC of 'dumbing down'.

Many newspapers use similar tactics to gain increased market share – for instance, the *Sun* usually has a competition on the front page because they know that this will attract their target audience to buy the paper.

PSB

The BBC is a different sort of organisation – a Public Service Broadcaster which is required to provide a public service when broadcasting and not to be driven by commercial influences. In other words, the BBC must provide a range of programmes for different social groups, including minority groups and niche audiences, without needing to justify this programming commercially. The BBC is funded by the licence fee which must be paid for owning or using a television set in this country rather than by advertising revenue.

Independent institutions

Independent institutions are often the hardest to categorise. However, in general, independent organisations are essentially commercial since they need to sell sufficient copies (e.g. of a magazine) or sell their programme to a TV channel for broadcast, for example. Their motivation may well be less commercial but most independents have to sell their texts somehow, in order to survive.

The jargon

Advertising revenue is the income received from advertisers who buy a slot for example, on ITV. They pay an agreed amount to show their advert at a particular time. Obviously, the more popular slots such as those during *Coronation Street* are the most expensive. This is why commercial institutions need to ensure they have big audiences.

Checkpoint 2

Investigate the viewing figures for the BBC channels for a single day. How large is the BBC Four audience? Looking at the schedule, do you think that BBC Four offers good value? What is the target audience for BBC Four?

You can find the viewing figures from BARB if your school or college subscribes (www.barb.co.uk) or from the BBC Press Office (www.bbc.co.uk/pressoffice) among other places.

Exam question answer: page 22

Investigate two magazines produced by the same institution in the same genre. (Look on the EMAP or Future Publishing website for lists of titles.) What institutional codes and conventions can you identify in each magazine? What seems to be the institutional context of production for these magazines?   (1 hour)

15

Narrative

Narrative basically means 'story'. All media texts have a narrative. It may be an explicit narrative in a soap opera or a film, say, or an implicit narrative, such as in an advert. Indeed, newspapers usually refer to a 'story' even though they are reporting fact, not fiction.

Identifying the narrative structure ●●●

This is a good starting point, as it will allow you to frame the rest of your comments appropriately. Once you have identified what type of narrative you are receiving, you should ask why you are receiving this information.

Hollywood screenwriters are taught that there is a very fixed formula for the narrative structure of a film. Quite simply, they are taught that the action should happen in three 'acts' and there should be a crisis point at the end of Act 1 and Act 2 and a happy resolution at the end of the film. Many narratives do not conform to this structure but it is important to remember that it is one we are used to from the fairy tales and myths which we first heard when we were very young. When studying a text it is reasonable to ask the following questions about the narrative structure:

➜ Has the institution created a text with a conventional narrative structure?
➜ If not, which conventions have they subverted or broken?
➜ Why have they done this?
➜ What impact does this have on the audience?
➜ What audience expectations have they established or challenged by using this structure?

In the case of an interactive text, the narrative may be less clear. Not only may the structure of a website be much more flexible, but the developing production and distribution of interactive movies will affect our cultural expectations of narratives. In an interactive movie, the narrative is directly controlled by the audience, which chooses its own path and establishes its own enigmas and crisis points. Each time it views the movie, it may experience it differently.

Computer games depend on audience expectations of narrative as well as good understanding of generic codes and conventions. The player must be aware of what is expected of them at each stage in the game if they are to be able to progress through the game. Many of the most sophisticated games prequel each level of the game with a brief narrative to give a semblance of a linear structure to support this. Do you find this helpful or do you prefer to create your own narrative when playing games?

Action point

How many different plots do you think there are? List those you can think of. Scriptwriters are told that there are only seven plots in the world, based on Greek mythology. So all media texts are reworking the same stories! Look at www.filmeducation.org/secondary/studyGuides/screenploy.pdf

Example

24
The US television serial *24* uses time in a different way from many programmes. Each serial takes place over 24 hours and the serial is broken up into 24 episodes, each of which is shown over an hour. There is always a clock in the corner of the screen showing in real time (for the programme) what the time is. The programme then uses a variety of narrative constructs such as parallel narratives in split screens to show events in a variety of locations but maintaining the real time of all these events, even using multiple frames on screen to show multiple narratives.

Plotting the events and actions ●●●

An **action** is something that a character does to himself, an object, or another character. An **event** is something that happens to the character, something that impacts on him/her, and over which he/she has little control. Being hit by a car is an event, but the antagonist intentionally hitting another character is an action. A narrative is usually a series of events and actions and the audience's expectations at any stage are often related to the pattern of events and actions taking place.

If a character in a TV programme appears to have no control over events and is a victim, it is less likely that an audience will be sympathetic to them. Where the character does things (i.e. actions) then the audience is able to engage with the character and the narrative becomes more emotive. This can sometimes be the difference between, say, a successful and a less successful comedy programme.

Time ●●●

Time is used in different ways in a moving image text – for instance, there is often a differential between **discourse** time and **story** time. **Discourse** time is the time taken to narrate the events. **Story** time refers to the 'real time' of the events. This can be established in a variety of ways:

→ **Summary** – In this case, discourse time is shorter than story time. A common way of establishing this is to use a fade on a changing clock face.

→ **Ellipsis** – This is the most common way of establishing discourse time rather than story time. If someone is shown getting out of bed, followed by a shot which shows them driving to work, we assume all the events in between, keeping the narrative moving.

→ **Scene** – Here, story time and discourse time are equal. Reality TV programmes and soap operas will often try to suggest story time even when it is not being used, to try and maintain a **realist** convention.

→ **Stretch** – Discourse time can be longer than story time, perhaps through use of slow-motion or a freeze frame. This can be a good way to directly affect an audience emotionally, for example.

→ **Flashbacks** change the linear progression of a text, moving away from a **realist** convention to allow the audience to judge or gain perspective on events.

Checkpoint 1

What is story time?

Checkpoint 2

What is discourse time?

The jargon

A *realist convention* means a text which is presented as if it is 'real' life – for example, a soap opera. A *non-realist convention* means a text which moves away from presenting reality – perhaps by a character speaking directly to camera, or a fantasy dream sequence.

Exam questions   answer: pages 22–23

Watch an episode of a TV programme which works mostly in story time, such as *24*, and one which works in discourse time, such as a typical police drama. In what ways is time established in these different narratives? How does this affect audience expectations and responses?   (1 hour)

Genre

Another of the key concepts essential to your study of the media is genre. At the most basic level, genre simply means 'type'. In this way we might talk about the genre of a television show being docudrama or the genre of a film being action/adventure. You will probably already be familiar with these words from your media consumption and general discussions. So why is genre a key concept?

Genre classification ●●●

If we start by thinking about television genres, programmes within the same genre will have similar, familiar or instantly recognisable patterns, techniques or conventions that include one or more of the following: setting, mise-en-scène, plot, motifs, technical features, situations, characters and format. For example, a soap opera, crime serial or reality TV programme sets up certain audience expectations as to what the programme will be about and how it will be constructed. Film genres can include action/adventure, sci-fi, Westerns, horror films and film noir.

When you are studying a text, it is helpful to start by identifying the various codes and conventions employed and relating these to the genre codes and conventions. This gives you a useful starting point for contextualising the text – in other words, remembering that no mass media text is made in isolation. Audiences apply their previous knowledge of each genre to create meaning for a text. This also works with print texts. For example, when we talk about 'niche market' magazines, what we really mean is magazines within a certain genre. So lifestyle magazines have a set of codes and conventions which we would expect to see – ranging from the image on the front cover, the lexis and representations to the content of the magazine itself.

All texts operate within a genre, although many texts are **hybrids** of more than one genre. For example, a development within documentary has been the docusoap which uses codes and conventions of documentary and of soap opera to create a new hybrid genre.

When *Loaded* magazine was launched, it managed to create a whole new genre, simply by rejecting many of the conventions of magazine publishing at the time. In this process it managed to create a new genre – the 'lads' magazine'.

Codes and conventions ●●●

Associated with any genre are its codes and conventions. This means the textual **codes** which give meaning to an audience and the **conventions** of the genre, such as themes/ideologies or narratives, which are used or subverted by the text.

These codes and conventions are important for the audience (because people know these codes and conventions are likely to be used in a text within this genre) and also for the institution because there is a genre framework for the text and it is more likely to attract audiences.

Checkpoint 1

Look at the TV listings for this evening and identify the genre of as many programmes as possible. Is this easy? Why?

The jargon

Niche magazines are ones which are published for a very particular, often very small audience. For example *Carp Monthly* is a niche magazine aimed at carp fishing hobbyists.

Example

When you are studying a slasher film, for example, you will probably be quickly able to identify a range of generic codes and conventions used in the text, such as knives and dark places.

Identifying genre

When analysing a text, it is always important to identify the genre of the text quickly, as this will help you to identify **what** particular codes and conventions are being used in the text and also help you analyse **why** they are being employed.

If you are analysing a film, the genre might be action/adventure, Western, sci-fi or musical, for example. If you are analysing a television programme, it might be sitcom, soap opera, children's drama or documentary. A print text might be a lifestyle magazine, a broadsheet newspaper or a film poster. A new media text might perhaps be a promotional CD-ROM, an e-commerce website or an FPS computer game. If you are analysing a tabloid newspaper, you will be expecting the conventional masthead, a dramatic headline, images of celebrities, competitions and particular types of content.

You should then be able to identify **how** and **where** some of these codes and conventions have been used in the text and therefore be able to write about **why** they have been used this way.

Audience ●●●

Genre is a useful 'hook' for an audience. Trailers for a new television serial will clearly identify the genre and, along with establishing an **enigma** which the audience will want to resolve, this genre identification attracts an audience. Just think about the number of similar programmes when any genre becomes popular with audiences: after the first *Big Brother* programme came many more reality TV shows and the success of *Changing Rooms* was followed by a range of property programmes.

So why do we like genres? ●●●

Given that there are many texts created within a particular genre, why do audiences continue to watch/read them? In many cases it is just this genre expectancy which an audience desires. It is possible, for example, to predict exactly what will happen at the end of a particular genre film. An audience will often find this predictability appealing. Look around the newsagents' and see how many people will buy more than one example of a magazine in a particular genre, such as gardening or film magazines. Audiences like genre because it enables them to pigeonhole a text and they feel secure interacting with a generic text.

<aside>

Checkpoint 2

What are generic conventions?

The jargon

FPS – First Person Shooter game where you take on the role of the main protagonist

</aside>

Exam question answer: page 23

Record episodes of two different television programmes within the same genre, for example, two hospital soaps or two property programmes. What generic codes and conventions can you identify in the two programmes? How does the use of these conventions fulfil audience expectations?   (1 hour)

Answers
Key concepts

Important note

To help you focus on the key concepts and think about how to write about them at this level, we have used a series of questions with a particular emphasis on each of the key concepts in turn. However, you should remember that when you are doing your Media Studies exams, it is likely that you will be given a far more general question and expected to use all the key concepts in your answer. The best answers will be able to write about the application of all the key concepts in some depth, not just identify that they are used.

Key concepts

Checkpoints

1 *Key concepts* are the main way in which we identify media texts, such as audience, institution and representation.
2 *A soap opera* is a programme such as *Eastenders*, *Coronation Street*, *Neighbours*, *Hollyoaks* or *Country Practice*.

Exam question

In this initial answer you should be making notes under the main MIGRAIN headings:
Media language
Institution
Genre
Representation
Audience
Ideology
Narrative
and writing down the media codes which you can identify under each heading to prove your point. Remember that each point must be supported by evidence from the text.

Once you have made some notes under each of these headings you should be able to draw lines between them to identify how they are linked. For example, you may find that you can make a point about the use of colour which relates to audience, media language, genre, ideology and representation. These are the sorts of points which are highly rewarded in exams because they show that you are aware of all the ways the text communicates with the audience.

Media language

Checkpoints

1 Some institutions will have a lot of control over the texts they construct (such as an independent film maker) but others may have limited control (e.g. where the audience are voting out characters from a reality game show).
2 You are most likely to feel that reality TV texts are negotiated between the institution (which has to define the context for the game, e.g. the jungle, and decide on initial participants) and an active audience, which controls the voting process and so shapes the later stages of the programme.

Exam question

This is a good example of an essay question with a ready made plan. Once you have selected your chosen music video, you can start by defining what the logos of the text might be and consider why such a text is unlikely to have a logos. To do this, you might want to write about the purpose of a music video as a promotional tool, promoting an image rather than a fully realised vision. Of course, there are some artists who seek to produce work which they see as having a logos and this could lead to some interesting discussion.

Once you have decided whether there is or is not a logos for the video, you can consider whether the audience passively receives the meaning constructed by the institution, negotiates meaning depending on their own cultural situation or actively creates a meaning for the text within their own cultural context. Your discussion of this might well develop into some commentary about music artists as 'institution' and the way that a music video relates to a target audience and to secondary audiences.

Representation

Checkpoints

1 To persuade an audience in a documentary, positive representations are used. So in this example, you would use representations of responsible 12-year olds, looking and sounding mature and promoting mature and responsible ideologies.
2 There are always stereotypes associated with a particular genre. Sci-fi, for example, suggests a charismatic but troubled leader, an intelligent and mature female sidekick, the maverick engineer and the warmongering alien.

Exam question

The usual social groups we study when writing about representation are grouped by age, class, religion, gender and disability. You might start by taking each of these groups in turn and looking at the ways they are represented in the bulletin – for instance, are they represented positively or negatively?

To analyse these representations you would look at the visual representations – what they are wearing and the mise-en-scène, for example, as well as technical codes like the use of camera angles or framing to create meaning. You should also write about aural signs such as whether they speak for themselves or whether someone provides a voiceover, perhaps controlling our view of this group? What words are used to describe the group or by the group? For example, a report about an attack on a pensioner may well use words such as 'helpless victim' and 'vulnerable' to establish an emotive response in the audience, which therefore represents the group 'pensioners' in a particular way.

When you are writing about all the representations used, remember to write about them in relation to the other key concepts as well, not just what the representation is and what media language is used to establish it. You should write about what institution made the bulletin (is it perhaps a national BBC bulletin – which implies a particular ideology) and how that affects the representations constructed. You should also consider the representations in terms of the target audience – do they match audience expectations, for example?

Audience

Checkpoints

1 By becoming aware of your own media consumption you are better able to identify how media producers create texts to appeal to particular audiences and the devices they employ. This awareness will also help you to become a more critical consumer and to develop critical autonomy.
2 Local newspapers have to target their audience very carefully. This is done through content (local stories), visuals (lots of pictures of local children) as well as discourse (stories about 'louts' causing trouble, for instance). Because their stories are primarily soft news presented as hard news, they will use emotive pictures and language to engage the audience.

Exam answer

If you are going to analyse the audiences across a range of channels and schedule points, it makes sense to break this down first. For example, can you define a primary target audience for each of the terrestrial channels and for some of the digital channels? (N.B. Why can it be hard to define the target audience for a channel such as the Documentary Channel?) You can also look through the listings to see how the target audience for each channel varies throughout the day.

Once you have done this basic analysis, you can start to draw some conclusions about the target audiences. Are there any clear patterns about when a group of channels all target a particular audience? Are there any clear signals on some of the more specialist channels about attracting a target audience at certain times of day? Why do you think some channels only broadcast at certain times? What does this suggest about their target audience?

From here you can move on to discuss the ways in which different channels manage to create an identity which attracts a particular target audience and maybe to think about the ways in which these channels seek to construct an audience who will therefore choose to watch the channel.

You might also want to think about whether there is competition in the way different channels target audiences at certain times of day. For example, does Sky try to attract a particular audience away from ITV at any point during the day, or does BBC Four compete directly with BBC in the evenings?

Do you think this suggests that there is a finite audience for all these channels or is the audience growing with the number of channels? Why do channels still vary their output over the course of a day, given that most audience members have access to VCRs and maybe more sophisticated ways of recording programmes?

Ideology

Checkpoints

1 The discourse used in news bulletins is one of the most important ways that the ideologies are conveyed to the audience (e.g. the use of the verb 'demand' rather than 'ask' gives a very different impression of the way a question is being presented). A good example from news media is the use of the term 'freedom fighters' as opposed to 'terrorists'.
2 Patriotism for an international sporting event is established by use of colours associated with the UK flag; long lead-ins to the event with interviews, pundits, etc and shots of excited fans. Less explicit means are also used, such as the use of cheering fans as a soundtrack underneath commentators even before the event.

Exam question

Obviously, as with other answers in this section, your answer will partly depend on the magazines you choose.

Your analysis is likely to start by identifying the target audience for each magazine and therefore the niche market. You should then identify the dominant ideologies for each text and relate these to the target audience and niche institutional context. Once you have identified these ideologies you can focus in detail on the ways the magazines construct these into aspirations based on these ideologies (e.g. I should use this diet in the magazine to make me more like the ideal person shown on the front cover). You may well consider the ways in which these magazine publishers create

these aspirations in society, e.g. the way they use models to construct a particular perception of female beauty.

It's a good idea to choose two magazines here which are for very different niche markets because it enables you to explore how the relevant ideologies are promoted for that target audience. A lifestyle magazine targeting men may promote ideologies relating to stereotypical male aspirations such as success at work or doing DIY work, while a magazine targeted at young women may well focus on stereotypical aspirations such as 'having the perfect haircut' or 'finding the perfect jeans'.

If you choose a mainstream and a more subversive magazine, you may be able to write about very contrasting ideologies and consider how the dominant ideologies are maintained in society and how some magazines are used by particular groups to try and subvert these ideologies.

Institution

Checkpoints

1 Because the *Blair Witch Project* was independent, they were able to use a range of techniques to make it convincing. For example, because the actors were not well known, the producers were able to imply that they were real characters and the creation of the *Blair Witch Project* website convinced many people that it was real. There was no explicit marketing campaign for the film, again heightening the sense that this was a real event.
2 BBC Four has very low viewing figures and would probably not be sustainable if it were a commercial digital channel. However, it is publicly funded and therefore supported because of the PSB charter provision from the BBC. The target audience is also the core audience of affluent professionals which the BBC seeks to retain.

Exam answer

As with other questions here, your analysis should start by making notes about the specific media language used in each magazine. You should be able to use the front cover, contents page and one or two spreads (two pages) from the magazine to do a detailed deconstruction for your answer. Do not use advertising pages because the adverts are constructed by a different institution and the advertisers simply buy the space to put the advert into the magazine.

When trying to identify similarities between the two magazines, you will probably start by looking at the media language:
- Is the layout of pages the same for each magazine?
- Are the same fonts used?
- Is the structure of the magazine the same?
- Are the same colours used?
- If you look at the list of writers, do the same people write for both magazines?
- What sort of representations are used in each magazine?
- Is the lexis (discourse) similar for each magazine?
- Are images constructed and used in the same way in each magazine?

A strong answer would then move beyond this to explore how house style can be used to create a wider audience, who possibly read more than one magazine from this publisher, perhaps reassured by familiarity of format. It would also cover the institutional practicalities of being able to produce many titles more easily if all conform to a group identity. Such answers might also go on to explore how ideologies can be established across differing titles in this way – even across different market sectors.

Less secure answers would probably respond on a more simplistic level, tending to directly compare and contrast the media language used in relation to target audience and to show the similarities across titles. This can often be a key differentiator for essays of this kind – the really interesting stuff comes with the second part of the question, which asks about institutional context because this is where you need to write about why a publisher will use such similar codes and conventions across a range of publications not just what these similarities and differences might be.

Narrative

Checkpoints

1 **Story time** is the sequence of events and the length of time that passes in the text. It is sometimes called 'real time' narrative.
2 **Discourse time**, on the other hand, covers the length of time that is taken up by the text and the sequence of events as they are presented in discourse – which may not be the sequence in which they happened.

Exam answer

When comparing time in narrative structures, we normally examine the differences in use of **real time**, **summary time**, **stretch**, **ellipsis**, **pause** and **sequencing** of events.

Story time/real time is obviously used throughout texts such as *24* which are presented as if in real time. Live news reports happen in real time, for example.

Summary time is used in conventional texts, e.g. when a graphic appears saying, 'the next day'. We don't need to see the intervening period as we only see the key moments.

Ellipsis time is similar – we see key events in a sequence and fill in the rest for ourselves. We don't need to see the middle bit, which keeps the narrative moving quickly. For example, we will see the policeman leave the police station and then see him pull up outside someone's house.

Stretch time and pause are obviously when the narrative is slowed down – slow motion sequences or freeze frames are good examples of these.

The last key area to analyse is 'sequence' time. In a real time programme such as *24* we should see events as they happen. In a conventional crime series, however, we may well have a narrative structure which shows us the end (e.g. the criminal being caught) before the main narrative reaches resolution, to control the way the audience engages with the text.

As with other answers in this question, your response must be structured in relation to the texts which you have chosen but you should cover at least some of these key devices for structuring narrative time, making good use of specific examples to structure your answer.

Genre

Checkpoints

1 It is usually very easy to identify the genre of a programme from the listings, either because the listings magazine identifies it or because it is obvious from the name of the programme. It is important to programme makers that the genre is apparent since they know this will attract audiences.

2 Generic conventions are the signs (such as plot, visual representations, generic props or stereotypical characters) associated with a particular genre.

Exam answer

An answer to this question should first identify typical codes and conventions in each programme which help to define the genre and fulfil audience expectations. The logical starting point is to look at those codes and conventions which are common to both programmes as they are likely to be a good starting point to evidence the importance of using these generic codes and conventions to satisfy audience expectations of the genre.

Revision checklist
Key concepts

1	Understand the key concepts used in Media Studies and how they work together	Confident	Not confident. **Revise** pages 4–5
2	Understand how to deconstruct the media language used in a media text	Confident	Not confident. **Revise** pages 6–7
3	Understand the way that an audience may receive and interpret a text	Confident	Not confident. **Revise** pages 10–11
4	Understand the concept of 'target audience' for a media text and how this influences the text	Confident	Not confident. **Revise** pages 10–11
5	Understand how an understanding of the target audience for a media text relates to the other key concepts	Confident	Not confident. **Revise** pages 10–11
6	Understand how representations of people and places are used in a media text	Confident	Not confident. **Revise** pages 8–9
7	Understand how these representations underpin our deconstruction of all the key concepts	Confident	Not confident. **Revise** pages 8–9
8	Identify the institution which has made a media text and how this affects the text	Confident	Not confident. **Revise** pages 14–15
9	Identify what basic ideologies may be operating in a text	Confident	Not confident. **Revise** pages 12–13
10	Identify the genre of a media text and the generic codes and conventions within it	Confident	Not confident. **Revise** pages 18–19
11	Identify the narrative structure of a media text	Confident	Not confident. **Revise** pages 16–17
12	Evaluate the use of time in a moving image text	Confident	Not confident. **Revise** page 17

Media language

All your work for AS and A2 level will centre on textual analysis because that is the starting point for all study of the media. Work in Media Studies depends on being able to 'read' media texts. This does not mean simply understanding what they are telling us but also being able to identify how they create meaning, how they present ideologies and representations of reality to us and how they present information to a particular target audience.

In this section we simply give you some tools to be able to write down your 'reading' of the text using media studies discourse (language) to show that you can deconstruct the text in detail and see how it creates and communicates meaning.

Although there are three different groups of text: moving image texts, print texts and interactive texts, you will see that you can read the key concepts in all these types of text and then explore features which are medium-dependent to create a very detailed reading of the text.

Exam themes

→ Key concepts
→ Texts in context
→ Deconstructing different types of media text

Topic checklist

○ AS ● A2	OCR	AQA	WJEC
Introduction to media language	○●	○●	○●
Audience	○●	○●	○●
Representation	○●	○●	○●
Semiotics	○●	○●	○●
Narrative and genre	○●	○●	○●
Lights, camera, action!	○●	○●	○●
Moving image	○●	○●	○●
Print texts	○●	○●	○●
Deconstructing new media	○●	○●	○●
Texts in context	○●	○●	○●

Introduction to media language

Reading the media

Reading the media is at the heart of Media Studies, which is why, although we always talk about the key concepts being integrated, we then start with media language. The link between media language and the other key concepts is fundamental but until you can understand media language, you won't get far with reading media texts.

Reading a text

Remember that reading a **text** is a lot broader for media studies than simply reading the words. When we read a text, we are looking at all the content elements and how all these elements work together to create meaning. If you are reading a newspaper, the masthead of the newspaper front page will give you the name of the newspaper. If you simply look at the words, you find out the name of the newspaper, e.g. the *Sun*. However, you find out much more about the audience, ideology and institution by analysing **how** the masthead is constructed, e.g. in terms of the colours used, the font used, how big the name of the newspaper is, and what other images or text are also put into the masthead. This is what we mean by 'reading the text'.

Constructing reality

All media texts are constructed; someone has made decisions about how they should be constructed so that the form matches the content and with a particular target audience in mind. (A party political broadcast presented as a music video for a new garage band would be unlikely to appeal to a very large target audience!)

Mass media institutions produce texts for primarily commercial reasons – if a text is well constructed and well targeted, within an appropriate context, it will reach more consumers, which generates more money. The audiences consume the texts within their own cultural, social, historical and economic context. The target audience for the text should be able to identify with the ideology within the text. When deconstructing the text, you need to remember that it is important to read how the target audience would consume the text so that you can structure your analysis accordingly.

Deconstructing the text

The description of deconstructing a text is generally used in relation to a particular way of reading a text, called 'semiotics' (see pp. 32–33), but is a very good description of the process of reading a text. When you write an essay, you construct that essay by planning what you will include and how you will structure the essay, before you write it. The process of constructing a media text is just as careful.

The jargon

A *masthead* is the section of the newspaper front page which holds the name of the newspaper and associated graphics or text. It is usually very easily identified and acts as a useful recognition device in a newsagent's.

Example

If you were writing about the use of a particular colour in a text, you might write about how the **media language** of the colour had particular associations (or connotations) for the target **audience** and was a typical **genre** convention, perhaps **representing** a particular person in a particular way, which suggested the **ideology** of the media text as being X for example. This might also be typical of the **institution**, which might have constructed the text in line with typical **narrative** conventions.

Checkpoint 1

Try looking again at the front cover of one of your textbooks. What can you say about the media language it uses and its audience?

Remember that, although you might assume that you only need to deconstruct texts when doing critical analysis, this is actually the key skill for your whole course. You need to be able to deconstruct your own projects to justify your design decisions, as part of the evaluation process. You need to be able to deconstruct texts when writing about genres, movements, auteurs or developments in media technology.

Remember – all media texts have been:

→ Carefully designed (i.e. someone has made decisions about what colours, layout, etc. to use).

→ Created for a particular context (a news programme aimed at children will be constructed very differently from a late night financial news bulletin, for example).

→ Created within a particular economic, social, cultural and historical framework. For instance, a low budget film will be very different from a Hollywood blockbuster, an underground Armenian film will have a different ideology from a mainstream Australian film and a modern remake of an old film will have access to very different technology.

When deconstructing a text, our job is to identify the different elements which construct the text and the contextual factors which frame it.

At AS level, your focus should be on the **elements of construction** – media language, representation and audience, to construct an **ideology** for the text, and showing understanding of how the **institutional** context controls and dictates production.

At A2 level, your approach should be based much more on the idea of text in **context**, i.e. the context of **production** and **consumption**. Therefore you need to show a greater depth of understanding of how the institutions control and shape texts and to be able to contextualise the text in terms of how the social, cultural, economic and historical contexts have affected the construction of the text.

In this section, we have started by thinking about how you might deconstruct an image, gone on to look at deconstructing print texts and then look in turn at deconstructing moving image, radio and lastly new media texts. All of these start from the same principle.

Checkpoint 2

Watch the opening sequence of either *Hollyoaks* or *Neighbours*. What do the clothes the characters wear suggest about them?

Watch out!

Each of the exam boards has a different way of examining textual analysis. Always check with your teacher for the exact requirements for you!

Don't forget

There are many different 'formulas' for how to deconstruct a text. It may be that you know one which you find helpful. USE IT! Some general suggestions are given here but there are many, many ways to deconstruct a text. The most important thing of all to remember is that the most important question to keep answering is always: WHY did they do that? Don't fall into the trap of simply identifying WHAT has been constructed.

Exam question   answer: page 46

Watch the teaser trailer for *Die Another Day* (2002). You can find this on the DVD edition of the film or through www.imdb.com.

Write a textual analysis of this teaser trailer.   (1 hour)

Audience

Consideration of the target audience is a vital part of the analysis of any text. You can look at the media language employed, identify the institution and ideology, consider the representations employed and the narrative structure and genre of the text but unless you match all that against the target audience for the text you are not doing a successful analysis.

Identifying the target audience

All mass media texts are made for a particular audience. Most of the time we take this for granted – we know that we are likely to enjoy some texts but be very uninterested in other texts. We know that we like some radio stations but would not listen to other stations.

Analysing a media text is much easier once you have identified the target audience, because then you can start to look for particular features of the text which appeal to that audience.

How to describe the target audience

The term 'target audience' for media producers means 'the audience we think we are making this product for'. Media producers identify the target audience during the planning stage for a media text, because it affects the shape of the text so much. That is also why you must decide what the target audience for a text is before you can really begin to analyse it in detail. For example:

→ Is the text made for children or for retired people?
→ Is it for a male or a female audience?
→ Is it for an existing audience or is it seeking to create a new audience?
→ Is it targeted at people living in a particular part of the country?
→ Is it targeted at people in a particular income bracket?
→ Is it for a particular type of audience such as aspirers or homemakers?
→ Is it for a narrow audience or a broad audience?

All these different target audiences will **mediate** the way a text is constructed because it will be shaped differently, structured differently and have different content depending on the target audience. By identifying the target audience, you make it far easier to say **why** a text has been constructed this way, rather than simply writing about **how** the text has been constructed.

Checkpoint 1

Why do you think the BBC has four different mainstream radio stations? What is the audience for each of these?

The jargon

Advertisers in particular define audiences by lifestyle as well as in other ways. *Aspirers* and *homemakers* are two of these categories. See pp. 98–99 for more information.

Checkpoint 2

What do you understand by the term 'mediate'? How does an audience mediate a text?

Audience theory ●●●

It may be that you also want to analyse your text in terms of audience theory. You may want to write about the **uses and gratifications** which the target audience gets from the text, for example, or to write about how the text may **affect** an audience. We have written more about audience theory in a different section of this book (pages 86–91) so you may find it helpful to look there to see how you might apply this when analysing a text. Even if you want to do this, you should remember that you must start by identifying what the target audience is before you can start to think about whether it is influenced by the text or why it chooses to engage with it.

Writing about audience ●●●

When analysing a text in an exam situation, it can be helpful to think about the audience for the text and how it is received by asking yourself a series of questions, such as:

➜ Who is the text aimed at (i.e. what is the target audience)?
➜ What assumptions are made about the audience which are revealed in the text's scheduling or positioning?
➜ Where and when is the audience likely to receive the text?
➜ How does this influence the form and structure of the text?
➜ How will this audience 'read' this text?

Reading a media text ●●●

David Morley undertook a study in 1980 to learn about how different target audiences received the same text. He looked at how different audiences received an edition of *Nationwide*, which was a local news programme shown after the main evening news on BBC One. He found that there tended to be three groups of readings of any text.

➜ A **preferred** reading is the reading most likely to be received by the target audience. This preferred reading is the reading which the media producers want the audience to receive.
➜ An **oppositional** reading is a reading by audience members who are not part of the target audience. They reject the preferred reading, receiving an alternative meaning from the text.
➜ A **negotiated** reading is one where the audience basically accepts the meaning the media producers intend for the text but it modifies the way it reads the text to suit its own position.

The jargon

Uses and gratifications and *effects* theory are the two types of audience theory which you are most likely to study at this level. You can find out more on pp. 90–91.

Links

You can find out more about 'reading' a text in the 'Media Language' section of this book, pp. 6–7.

Checkpoint 3

Think of the television programmes which you watched yesterday. Did you receive a preferred, negotiated or oppositional reading for each of these? If you were watching with other people, did you all read the texts the same way?

Exam question answer: page 46

Look at the front cover of this month's copy of a men's magazine. In what ways does this text target its audience? How might different audiences read this text?   (1 hour)

Representation

When analysing a media text, one of the most important things you need to do is to identify the representations used by the text and write about these in relation to the target audience through the use of media language. All the representations in the text are there to establish and confirm the ideology constructed by the media language. This is why your teachers may well talk about the representations in a text when they are analysing the ideologies behind a text as these representations are usually one of the best ways of being able to identify that ideology.

What is being represented?

Given that all media texts represent reality in some way, it is very important to write about the representations within a text, especially when writing about how the text appeals to a **target audience** and how it establishes its **ideology**.

You may need to write about different types of representation. Generally, we expect to write about representations of people and/or place and consider the reasons why they have been represented in this way in a text.

Questions you might ask:

→ Who or what is being represented?
→ Is this a positive or a negative representation?
→ How is it being established (visually, aurally, juxtaposition against mise-en-scène, use of camera . . .)?
→ How does this representation relate to the target audience?
→ What ideologies are suggested by this representation?
→ Is it a stereotypical representation?
→ Is it a fair and/or accurate representation?
→ Are the representations conventional or subversive?
→ Are there representations which you would have expected to see which are not used in the text?

Remember that when you are writing your answer, it is important to explain **why** the producers of the text have used these representations in the text. Are they seeking to appeal to the target audience or to alienate the target audience? Representations are a very powerful way of conveying a lot of information to the audience so it is very important that you can write about them while you are doing your analysis.

Representations of people are the most common types of representation to analyse. You might be looking at representations of individuals or groups. For example, you might have an advertisement using a representation of a 'typical' working mum to sell a product, while another advert might use representations of a group such as commuters to sell something else.

As we discussed in the key concepts section, when you are writing about representations of people in a media text, you are probably going to write about the representation of them in terms of:

Checkpoint 1

Think about how students are represented in your school brochures for new parents. In what ways are these representations used to suggest that the school/college is a happy place?

Links

Look back at the pages about representation in the first section of this book pages 8–9 if you are unsure about what we mean by representation.

- → age
- → gender
- → race
- → disability
- → socio-economic status
- → nationality
- → sexuality.

Usually we analyse representation by looking first at visual signs, because we absorb a great deal of visual information in a very short space of time and this is useful for media producers. But audio cues can also be important – think, for example, of the music used in a horror movie when the 'baddie' first appears but seems normal. This representation quickly tells the audience that this is the baddie and thus frames their responses to this character.

Representations of groups often become stereotypes. For example, representations of nations, e.g. 'the French' or 'the Irish', carry a lot of connotations within the stereotype which an audience is expected to be able to tap into when receiving the text. Gender representations can also be stereotypical – and sometimes offensive!

Representations of place can be stereotypical as well. The first time that British people visit New York, they often say 'It's just as I expected', because they have seen New York on TV and in films so often they feel very familiar with the place. These sorts of representations are usually culturally constructed as well – the representation of New York in a mainstream British soap opera is going to be very different from the culturally constructed representation of New York used in a news programme in Beijing!

Fiction and non-fiction texts ●●●

Remember that representation is not just an issue for fictional texts such as action/adventure movies. Many people complain to the Press Complaints Commission about the way they have been represented in a newspaper story. Whatever media text you are studying, there will be representations of a reality being constructed and your job is to identify **what** these are (**media language**) and **why** they are being used by the media producers. What **ideology** is being constructed and how are these elements interacting with the **target audience**?

Checkpoint 2

Think about how people of different ages/genders are represented on TV. Analyse the representation of age and gender for the programmes shown on a single terrestrial channel (e.g. BBC One) over a day. What does this tell you about changing target audiences?

Checkpoint 3

Representations of New York have obviously changed since 9/11. In what ways is New York represented now?

Links

Look at the visual representations of New York on the Ground Zero website (http://groundzero.nyc.ny.us/). What ideologies are being constructed here?

Exam question   answer: pages 46–47

Watch an edition of a lifestyle documentary such as a property programme. What representations are being used here? Why? You may wish to look at representations of people or of place or both.   (1 hour)

Semiotics

Semiotics enables us to deconstruct how **visual** messages as well as textual messages communicate. Semiotics is a 'science of signs'. By deconstructing a text using semiotics we have a tool which helps us to identify all the different elements of a text and see how they are communicating with the audience.

We generally deconstruct texts at three levels. These are: syntactic, representational and symbolic.

Syntactic level

This is the most basic level of analysis – it identifies the **denotations** of the text (i.e. it describes **what** is in the text). To look at a page/document, for example, at the syntactic level, hold it far enough away so that you cannot read the words and look at the purely visual message that the page/document communicates. Essentially, you are describing the text.

So, when you are deconstructing a text at this level, you focus on the immediate impression.

→ What are the dominant elements on the page/screen?
→ How is colour being used?
→ What is the overall effect of the text?

Representational level

The representational level examines the **meaning** of something. At this level you should look at the key concepts, ideas and objects included in the text and the **relationship** between them – has camera angle or font size been used to convey a difference in status for different characters or sections of text, for example? At this level you are thinking about how the text creates meanings through the **linking** of elements in the layout, the editing, etc. You are looking at how media language and representations are used to convey meaning to the audience.

Symbolic level

This third level is the **symbolic** level. At this level you are looking at what hidden meanings or connotations the text carries in terms of **ideology** and **institution**.

It is important to remember that a text is also dependent on the **context of consumption** – connotations will depend on the readers' own **cultural history**. For example, some symbolic signs can be interpreted differently in different contexts. In this way you can identify the **connotations** of the text and the **cultural context** in which it was created, which will help you to explore the relationship between the text and the **target audience** – a very useful skill when doing any form of critical analysis.

Checkpoint 1

Look at a range of women's magazines for a particular month/week. In what ways do the colours and images identify the season in which the magazines are printed?

Examiner's secrets

Examiners don't expect you to be able to use all these terms in your work but it is useful to be able to use a few – for instance **connotation** is a useful word to identify the ideology of a text.

Check the net

Daniel Chandler gives a very detailed blueprint for undergraduate students doing a semiotic analysis, which you can find at http://www.aber.ac.uk/media/Documents/S4B/sem12.html.

Sign systems

A sign system has two elements – a **signified** and a **signifier**. The 'thing' which is represented by the sign is the signifer and the 'concept' represented in the sign is the signified. Of course, many signs will be **polysemic** in that they will carry lots of signifieds from the one sign – e.g. the signifier of a knife on a horror poster has many signifieds associated with it.

Types of sign

C S Peirce suggested that there were different sorts of signs and that these types of sign should be treated differently. The difference comes in the scale of difference between signifier and signified.

→ The closest relationship is an **index** sign. Indexical signs are very closely related to the concept they signify – a tin of catfood signifies the catfood inside.

→ **Iconic** signs are **like** the signified. A photograph is often described as an iconic sign because it is like the person but is clearly separated from them.

→ **Symbolic** signs do not have an obvious relationship with the signified – a dove signifying peace is an example of a symbolic sign.

→ **Arbitary** signs are signs for which the meaning can change. Language is often an arbitary sign – there is no clear connection between the signifier and the signified. Current words which illustrate this concept include words such as 'minging' and 'bling', which carry particular connotations with particular groups, yet are meaningless in themselves.

You might recognise some of the signs here –

→ Which ones are indexical, iconic, symbolic or even arbitrary?

→ How many of these are now so accepted that we don't even stop to think what they are?

→ How would you explain them to someone who had never used this program or to someone who had never used a computer?

Sign systems can be very useful to convey a lot of information very simply. In media texts the sign system conveys many layers of meaning at the same time.

Exam question   answer: page 47

Provide a semiotic reading of this week's copy of a television listings magazine. Identify each of the four levels of sign (indexical, iconic, symbolic and arbitrary, and explain how they operate together to communicate meaning to the reader.   (1 hour)

The jargon

Polysemic means 'carrying lots of meanings at the same time'.

Checkpoint 2

If you see a poster for a new film which uses an image of a gun covered in something red, what might this signify?

Checkpoint 3

Look at a typical webpage, such as the BBC home page. Which visual signs on there are indexical, which iconic and which symbolic?

Examiner's secrets

It can be useful to be able to identify particular types of sign in a text but don't worry about getting it 'wrong'. You are very unlikely ever to lose marks by identifying the type of sign incorrectly – you are far more likely to pick up marks for engaging well with the text!

Narrative and genre

Narrative as a key concept is looked at further on in this book (see pp. 92–93) but as with all concepts in Media Studies, it is impossible to separate these concepts out. Thinking about the narrative being established by a media text is a fundamental way of identifying what the text is about and why it has been constructed that way – in other words, the ideology being established.

How can a print text have a narrative? ●●●

All media texts tell a story in some way – for instance, a film poster is intended to set up an **enigma** for you to resolve by watching the film. An advert is designed to make you want the product which is being advertised. A newspaper seeks to give an account of news events – but in line with its own ideology and political position.

Deconstructing the narrative ●●●

When thinking about narrative in any form of media text, the questions you might want to ask yourself include:

→ Is this a whole narrative, a section or an enigma?
→ If you are looking at a section of a narrative, what function does this section have for the narrative?
→ Who/what are the main narrative agents for this text? How do we know?
→ Does it conform to a formula?

Once you have identified the form of the narrative, you might want to look at the structure in more detail. You might consider the following questions and ask yourself why the narrative is structured in this way.

→ Is it linear or is the narrative being presented in some other way, such as with flashbacks, flash-forwards, or episodically?
→ Is it being presented in segments, separated by changes in location or time?
→ Is it being broken up or linked by graphics, say?
→ Is there a single narrative (even if different aspects are being shown) or are multiple narratives being used?
→ Is a clear narrative context being established? Many documentaries use a 'voice of God' voiceover to structure the documentary, for example, defining the text's ideology by giving this perspective on the action being presented.
→ Are there clear character roles and functions being established, such as the hero or the villain?
→ In what ways is the narrative constructing the ideology of the text? Is this overt, as in many adverts or covert, as in an action/adventure movie?

The jargon

Enigma means 'mystery to be solved' and is a term used by Barthes in his theory of narrative. Look at p. 93.

Checkpoint 1

Look through a typical Sunday supplement magazine. Which adverts can you readily identify a 'story' for? What is the narrative in the other adverts?

Example

Branigan (*Narrative Comprehension and Film*, Routledge 1992, p. 14) describes an outline of how any narrative will work:
1 introduction of setting and characters;
2 explanation of a state of affairs;
3 initiating event;
4 emotional response or statement of a goal by the protagonist;
5 complicating actions;
6 outcome;
7 reactions to the outcome.

The jargon

A *realist convention* refers to trying to imply 'real life' and using a very 'normal' style of presentation.

Checkpoint 2

Watch the OS of a typical Hollywood action/adventure movie. In what ways do story time and discourse time differ? Why?

Genre

● ● ●

Whenever we receive a media text, our responses to that text are framed by the genre of the text. We expect certain things of the text, depending on this genre. If we are receiving an action/adventure film, for example, we expect certain things to happen in the film. If we are receiving a broadsheet newspaper, we have expectations about the content and form of the text.

Codes and conventions

Associated with any genre are its codes and conventions. This means the textual **codes** which give meaning to an audience and the **conventions** of the genre, such as themes/ideologies or narratives, which are used or subverted by the text. These codes and conventions are important for the audience (because they know they are likely to be used in a text within this genre) and also for the institution because there is a genre framework for the text and it is more likely to attract audiences.

Identifying genre

When analysing a text, it is always important to identify the genre of the text quickly, as this will help you to identify **what** particular codes and conventions are being used in the text, and to analyse **why** they are being employed.

If you are analysing a film, the genre might be action/adventure, Western, sci-fi or musical, for example. In the case of a television programme, it might be sitcom, soap opera, children's drama or documentary. A print text might be a lifestyle magazine, a broadsheet newspaper or a film poster, while a new media text might be a promotional CD-ROM, an e-commerce website or a FPS computer game.

Analysing genre

The first thing you need to do is mentally list all the codes and conventions audiences might **expect** to see. So if you are analysing a Western, you would expect to see cowboys, guns, horses, deserts, etc. If you are analysing a tabloid newspaper, you will be expecting the conventional masthead, a dramatic headline, images of celebrities, competitions and particular types of content.

You should then be able to identify **how** and **where** some of these codes and conventions have been used in the text and therefore be able to write about **why** they have been used this way to appeal to the target audience.

Checkpoint 3

Look at the TV listings for this evening and identify the genre of as many programmes as possible. Is this easy? Why?

The jargon

FPS – First Person Shooter game where you take on the role of the main protagonist

---

Exam question   answer: pages 47–48

Find a copy of this week's issue of *The Lady* magazine or a similar ladies' magazine. What narratives are being established on the front cover? How does the front cover define the genre of the magazine?

Lights, camera, action!

The jargon

Mise-en-scène literally means 'put in the picture' and describes everything which is seen on the screen.

Checkpoint 1

If you watch the beginning of a classic film, such as *The Searchers* you will see that the mise-en-scène establishes a great deal about what is to happen in the film. How is the mise-en-scène used to achieve this?

Examiner's secrets

A very easy way to lose marks when writing about mise-en-scène is to **describe** the mise-en-scène without explaining **why** it is constructed this way!

When thinking about moving image texts, it is important to start your analysis by analysing what you can see and hear.

Mise-en-scène ●●●

The mise-en-scène includes everything which can be seen on screen. Analysing it is very much the same as deconstructing an image, as you saw earlier – just that the process is being applied to moving images rather than print images.

When writing about the mise-en-scène, it is sometimes helpful to break it down into different elements – think of S-C-A-L:

Setting:

→ Is the sequence set on location or is it filmed in a studio?
→ How is colour being used to convey meaning?
→ Is it realist (normal) or abstract (bizarre, e.g. a dream sequence) in form?
→ Is it expected for this text or unexpected?
→ What props or aspects of the setting have symbolic importance (e.g. a large knife to be used by one of the characters)?
→ What ideology is established by this setting?

Costume and make-up:

→ What representations are being constructed by the costumes/make-up and image of the characters/presenters?
→ What clues are given by the costumes about the characters'/presenters' job/status/behaviour?
→ How is colour being used?

Actors/presenters:

→ What are characters/presenters wearing?
→ How do they move/speak/appear, etc?
→ Are they open or closed to the camera?
→ Are they presented as belonging in this setting or contrasted to the setting?

Lighting:

→ What is shown in the light or in shadow?
→ Is the lighting harsh or soft?
→ Is it natural lighting or are special effects being used?
→ What is lighting being used to emphasise?

As before, each and every part of your analysis of the mise-en-scène must be followed by explanation – remember that the most important part of critical analysis is that you explain WHY something has been done!

Sound

●●●

Although most people take sound for granted in a media text, it is very important to establish context and atmosphere. Arguably, some of the best soundtracks are those which are so effective that we are hardly aware that they are doing this.

Sound is so important to our receiving of media texts that we take it for granted. We don't realise how important the sound is until we don't hear it. Sound tends to make a text more natural. **Non-diegetic** sound comes from outside the frame, such as the orchestra which often starts playing at a crucial moment in a romantic comedy. **Diegetic** sound is sound which comes from within the frame, such as a telephone ringing or the noise from a radio playing in the background of a scene. Another important attribute of sound is the way it works with, or against an image. **Contrapuntal** sound is opposed to or in conflict with the image; **parallel** sound matches with the image.

To get a good idea of the importance of sound to a text, simply watch a sequence from a typical horror movie. Try watching the sequence with the sound turned down and then watch it with the music again. You will realise how much influence the music has on your emotional engagement with the action.

Music

When considering the use of music in a sequence, begin by thinking about what type of music is being used and why and then think about whether it is familiar or unusual in some way.

→ Is the music contrapuntal or parallel to the action?
→ Is it diegetic or non-diegetic?
→ Is it reassuring or unsettling the audience?
→ What sort of atmosphere is it creating?

Sound effects

If sound effects are being used in the sequence, they should be important. Think about what sound effects are being used and whether the action would be as successful without them.

Language and silence

The language used by a character or presenter will usually reveal a lot about the representation constructed. Think about accent, tone of voice, vocabulary, conversation style and register. And don't forget that if a text is making use of silence, there is likely to be a very good reason for it!

Remember also that if the text makes use of a voiceover of some kind, you can analyse that voice and think about what it adds to the text.

Example

Theme songs – Perhaps one of the most familiar things about *Friends*, which has been running on E4 and Channel 4 for many years in this country now, is the title sequence. The combination of the very well known theme song and the very upbeat graphics immediately sets up audience expectations. Perhaps this is why the programme makers have changed the title sequence so little over the years. There are other programmes where the theme song is iconic of the text as well. Try looking on www.tv.cream.org for other examples.

Checkpoint 2

What sound do we hear at the end of every episode of *Eastenders*? Why is it used? Would a visual sign be as effective?

Exam question   answer: page 48

Watch the title sequence for the early evening news. In what ways do the mise-en-scène and the soundtrack establish our expectations for the programme to follow?   (1 hour)

Moving image

Example

Watch the pre-credit sequence of any of the *Terminator* movies. How are the audience's expectations for the rest of the movie set up?

The jargon

A *leitmotif* is a repeated symbol or sign associated with a particular character in the narrative.

Checkpoint 1

Record the opening sequence (OS) for the 6 o'clock news on BBC One. Make a list of each stage in the OS and identify why the sequence is constructed in this way.

It's very easy to take the camera work and editing for granted in a moving image text. It may be that the text that you are studying has obvious features in terms of camera work or editing which you can identify. Maybe it makes use of unusual camera angles or the editing is obviously non-naturalistic. Most of the time, however, the camera work and editing should be unobtrusive but will be fundamental to establishing the atmosphere. Your job is to become aware of what is happening and how this establishes the atmosphere.

Filming

When creating a moving image text, the camera can convey a lot of information to the audience. More importantly, it can control how we respond to the film. A panning shot at the beginning of a film shows us where the action will take place and therefore establishes the context for the action. A long shot of a run-down housing estate can convey a lot about the life led by the subject of a hard-hitting documentary. A close-up on the hero's face as the heroine walks away can convey his emotions to the audience without the need for words.

Some basic questions to think about while analysing the camera work and editing are given below.

Basic form
→ Has it been filmed in black and white or colour?
→ Is the use of colour symbolic?
→ Is colour/form used as leitmotif?
→ Is it in slow motion, accelerated motion, freeze frame or even time-lapse?
→ Is it filmed in wide-angle or telephoto lens or with a zoom lens?

Camera/Framing:
→ Is the camera high-angle, low-angle, eye-level, tilted?
→ Is it using extreme long shots, long shots, medium shots or (extreme) close-ups?
→ Is the camera static or does it use pans?
→ Does it use tracking or dolly shots?
→ Does the camera always follow the action?
→ Does it offer new perspectives on the narrative?
→ Is it subjective or objective camera work?
→ What type of shot is being used?

Shot types
There are several main types of shot used when filming, which range from close-ups to long-shots. These are the very basic types and their shortened forms;

→ CU = close-up (ECU = extreme close-up)
→ MS = mid shot
→ LS = long shot (ELS = extreme long shot).

Moving image

It's not enough to simply consider the mise-en-scène or the framing of a shot; you also need to think about the sequencing of the shots and how this adds meaning. For example, a series of very quick shots can create energy or suggest a lot of things happening. A slow pan around a room followed by a POV (point of view) shot tracking feet will immediately create tension and suggest someone is being followed. It's also a typical convention for a horror movie, yet unlikely in a romantic comedy. A news broadcast will usually be edited a particular way, to create a particular type of meaning – as with the CNN broadcasts during the Iraq war which we mentioned earlier.

Transition techniques

The most common type of edit between two shots is the cut – a 'jump cut' moves between two different sections of the text. Where a text is edited with some other kind of transition, for example a gradual change, such as a dissolve or a fade in/out, it will have been done for a reason and you should always ask yourself why. Think about how the editing creates or comments on relationships between characters and spaces. For instance, an easy way of suggesting a romance between two characters is to link them by editing.

Rhythm and pace

A fast pace creates a different type of tension from a slow pace, e.g. in a horror movie. And an uneven rhythm will cause anxiety in an audience, whereas a steady rhythm keeps them concentrating and reassured. Think about the beginning of many horror movies which start by showing 'normal life', with a steady, calm rhythm which is then destabilised by the villain.

Think about whether any unusual effects have been used. Although you might expect unusual transitions and strange camera angles in a music video, you would not expect them in a soap opera. Yet many texts use these effects at times to create meaning. *Eastenders* is a typical naturalistic soap shown on BBC One several times a week. Sometimes, though, the producers will make use of unusual camera angles or less conventional editing, in order to create meaning and keep the text fresh for a loyal audience. Does this make it any less naturalistic or is it important to use the camera creatively to establish meaning? Do you find it helpful when meaning is established in this sort of way or do you prefer your texts to remain more conventional? How can atmosphere and emotion be conveyed without using these techniques?

> **Checkpoint 2**
>
> What do the representations used in Channel 5 news bulletins suggest about the ideology of the bulletins?

> **Checkpoint 3**
>
> Watch the opening sequence of *Scream* and notice the ways in which the rhythm is disturbed. What effect does this have?

> **Example**
>
> Such non-realist techniques can be quite common – think about whenever anyone receives a letter in a soap opera. Almost always, there is a voiceover from the letter writer – does this matter?

> **Exam question**   answer: page 48
>
> Watch the pre-credit sequence and title sequence of *Matrix Reloaded*. In what ways does this sequence visually set up the audience's expectations for the rest of the film? You should think about form, camera angle, framing and rhythm and pace in the filming for your answer.   (1 hour)

Print texts

One of the most important things you need to be able to do in Media Studies is write and talk about media language – the elements which construct the texts.

What is media language? ●●●

As we said previously, media language is the construction of a text, using these elements.

For example, if you were constructing a new magazine, to be aimed at young people making decisions about where to apply for higher education courses, called UCAS-UCAN!, which of these fonts would you use for the masthead?

UCAS-UCAN!

UCAS-UCAN!

UCAS-UCAN!

UCAS-UCAN!

In what ways would each of these fonts change our expectations of the content of the magazine? How would they change our expectations of the target audience?

What expectations would be set up by different colours for the masthead? Would you choose a very pastel colour, for example, or a dynamic bold colour? Why?

What sort of image would you expect to use on the front cover of the magazine? Would you use a picture of a very famous university, a newer university, a typical student or perhaps a group of students? Why? What expectations would each picture set up about the form and content of the magazine?

Reading images ●●●

Of course, it is one thing being able to think in general terms about what would be a suitable image for a particular context but to analyse an image in depth you need to look a lot more closely. The old saying: 'A picture says more than a thousand words' is very true for all media texts. Our brains are tuned to relate rapidly to visual information and so a visual text can communicate with us very quickly. Once you start looking at images in more depth, you'll soon get into the habit of deconstructing them and realising how much care has been taken to make sure that the image communicates well with an audience. We take a lot of this information for granted, as when we make assumptions about what type of images we might expect on the cover of a niche magazine. Your job, when deconstructing texts, is to be able to write about all these assumptions we make. Media producers will have had

Checkpoint 1

Look at the masthead on a teen magazine. What does the masthead convey about the style, audience and content of the magazine?

"My way leads towards the creation of a fresh perception of the world. Thus I explain in a new way the world unknown to you."

Soviet film director Dziga Vertov (taken from *Ways of Seeing*, John Berger, Penguin 1990).

a visual construct in their minds when the text was constructed – you need to identify what that is and deconstruct the reasons why they made it that way.

When you are deconstructing an image, start by thinking about what the image is trying to communicate. What **ideology** is being established? How is this being communicated?

For example, look at this picture, which was intended to become part of an advert, advertising a new brand of toffee.

→ Where is the picture set? What can we work out about this location? (What clues tell us about the location?)
→ Is the person dressed suitably for this location?
→ What might be happening?
→ Does she seem to be happy? What might this suggest?
→ What ideology is being constructed here?

Example

Look at any picture showing a group of people. Think about **cropping** the picture so that you could only see one person. Find three different people in the image and draw a box around each one as if you were going to crop the picture down. Does this create different meanings from the original picture? What does this tell you about how much control a newspaper editor has when choosing pictures for a story for example?

Examiner's secrets

Remember that we expect you to be able to identify what is happening in a text which you are deconstructing. The key to getting good grades is to remember to write about WHY the text was constructed this way

Checkpoint 2

This is a black and white picture. If it were in colour, what colour would you choose for her suit? Why?

Exam question   answer: pages 48–49

Compare and contrast the front covers of *Woman's Own* and *Vogue* for the same period. In what ways does each identify a different target audience?

(1 hour)

Deconstructing new media

In many ways, deconstructing a new media text is likely to use the same principles and methods as deconstructing a print text or a moving image text. The key difference for a new media text will be interactivity, as this tends to be what differentiates a new media text.

Deconstructing a new media text

Genre and function

A logical starting point when deconstructing a new media text is to start by identifying the genre and function of the text. You might be looking at the first level of the latest version of *Half Life*, for example, or a website created to promote a new film. You might be looking at a training resource created in a particular context, or a weblog of some kind. Each of these will have a different context of production and context of consumption and you need to think about this as part of your deconstruction. Ask yourself:

→ What is the function of this text?
→ Is it entertainment?
→ Education?
→ Information?
→ Does it operate in isolation (like most films) or is it part of a larger network?

Institution

Once you have identified the form of text and started to think about the target audience, you should think carefully about the institution. A new media text can often be produced very cheaply and readily distributed via the Internet without needing anything like the financial investment required to launch a new film, for example. It may be that the text is produced by an amateur – as with a weblog – and in this case, the institutional values would obviously be very different.

Computer games, websites, CD-ROMs, WAP . . .

At the moment, the term 'new media' covers a wide range of interactive (and not so interactive) media forms. You should always start your deconstruction by identifying the basic features of the text in front of you:

→ What sort of media text is it?
→ Who was it made by?
→ When was it made?
→ What genre does it belong to?
→ Is it interactive?
→ What is the target audience?

After all, a Playstation FPS game is going to be made very differently from the official website for Honda Motorcars, for example.

Check the net

A resource which may be useful to you is www.webpagesthatsuck.com which compiles a list of bad websites and deconstructs what is so poor about each of these sites. You even get to deconstruct the sites yourself before reading what others have to say about the sites, giving you good practice for your work!

Checkpoint 1

Look at the website for a recently released film. In what ways does the site reflect the film? In what ways has the site tried to establish a separate identity? Why?

Links

You can find out more about computer game genres on pp. 136–137.

Multimedia texts

An analysis of an interactive text has to be different from an analysis of a static or linear text. Not only must you identify the different elements in the text but also how they interact and how the user interacts with them. You may find it helpful to start a new media analysis by thinking first about the form and function of the text in some detail and then analysing how this is constructed and why it is constructed in this way. In this way you will find it easier to identify the **genre** and **ideology** by identifying the **media language** and institution first and then the **target audience**. Once you have done that, you will be able to quantify the **representations** used and the **interaction** demanded in terms of audience engagement. A static or linear text cannot engage the audience in the same way, so operates much more straightforwardly.

Elements

When deconstructing elements of a new media text, you will need to think about the relationship established between the text and the consumer by the interactivity as well as the elements themselves. You also need to think about how these elements are used to interact with the user.

→ What visual elements are used?
→ Are they used in a functional or emotive way?
→ What forms of interactivity are used?
→ How is sound used by the text?
→ What support is offered to the user?
→ What codes and conventions from this media form have been employed?

Navigation

New media designers take a lot of time to think about the navigation for their text to make sure it is successful. Studies have shown that the most successful websites, for example, are those which are easy to navigate. When you are deconstructing a webpage you need to analyse this navigation and think about how effective it is. Of course, in some situations, such as strategy games, the navigation is deliberately obscure and complex as part of the gameplay involves working this out!

Checkpoint 2

What genres of computer game can you identify? What might the target audience be for each type?

Links

Remember that if you are not sure about any of these key concepts such as genre, you can look back to the key concepts section earlier in this book.

Exam question   answer: page 49

Either deconstruct four pages from www.bbc.co.uk (so use the home page and three other pages) or deconstruct the OS or a trailer for a mainstream computer game. How does this text establish audience expectations and genre?   (1 hour)

Texts in context

Action point

It's all too easy when doing a Media Studies course to look only at texts which appeal to you – which probably means texts for which you are in the target audience. Yet how can you really get a grip on a range of media texts if you only watch/read/listen to or interact with a limited range of texts?

Checkpoint 1

What different things do you say about a music programme when you are watching with your parents or your peers? How does this context affect your reading of the text?

Watch out!

At AS level your focus for analysis MUST be the text itself – the **way** it was constructed and the reasons for this construction. At A2 level you need to be much more aware of the **context** of production and consumption and how this influences both the construction and the consumption of the text.

Checkpoint 1

The **water-cooler effect** is a description of something which people will want to talk about at work the following day – traditionally while getting a glass of water. Advertisers obviously want to create this sort of effect for their adverts as this increases their potential buyers. Can you think of other genres or mediums where the makers of the text might be very keen to generate this effect? (Hint: Think about *Big Brother*!)

When you have been asked to analyse a moving image text, it is important to remember that it is a **mass media text** made for a specific audience. If you can **contextualise** your comments in this way while you are doing your analysis, you will find it very helpful.

Contexts of production and consumption

All mass media texts have two contexts that you need to think about when analysing them – the context in which they were **made** and the context in which they are **consumed**. You need to think about why the text was made that way – what influenced the construction, and why a certain audience will like it. We make certain assumptions, both methodological and theoretical ones. Even the statement: 'I didn't like this film' conveys a lot of assumptions and expectations about the film's content, form and ideology.

So, when studying a text, it is important to think about:

→ The **historical** context for the text: **economic**, **social** and **political** factors that influenced the way it was made.
→ The **genre** or form from which the text starts.
→ The **cultural context of construction** (the conventions, attitudes, expectations and ideologies which are part of this genre context).
→ The **institutional** context of the text.
→ **Other texts** created by the same institution, auteur or team, or within the same genre.
→ The **context of consumption** for the text. Is this a film which is designed for multiplexes or to be shown on video, for example?
→ Whether the text is using **intertextuality** in some way to communicate with the audience. Is there an expectation that the audience will be familiar with other texts? Why?

All these concepts are looked at in more detail in other sections of this book but you need to be aware of deconstructing a text into its context whenever you start to analyse it. If you are comparing texts, it is even more important to think about their contexts.

The following example illustrates the importance of context. Towards the end of the Iraq war, the Iraqi authorities ordered Reuters (the famous international news gathering agency) to stop distributing video material to CNN. Reuters was allowed to continue to send live video material to other news organisations but not to CNN because the Iraqi authorities felt that it was not being shown in **context** and was being used as US propaganda about the war rather than as news reporting. The **context of production** (i.e. the generating of news

reports using this footage by CNN) and the **context of consumption** (patriotic US families celebrating US victory) was felt to be so biased by the Iraqis that Reuters was told to stop giving the material to CNN or risk losing all permission to gather news in Iraq. It is perhaps a measure of how powerful the Iraqis perceived this context to be that it was only video material which was forbidden!

Institutional context

One of the most important contexts to think about then is the insitutional context – in other words:

➜ who made it
➜ why they made it
➜ when/where it was distributed
➜ assumptions made about target audience.

You will find it very hard to write in much detail about media texts unless you remember to think about **who** made them and how this has influenced the context of production and consumption. Most of the time we assume this knowledge – we expect to watch particular television channels, for example, and expect a particular ideological position from a certain newspaper. Yet when you are analysing texts, you need to be able to identify what the institution is and how this has affected the way the text was constructed:

➜ An action/adventure film made by a mainstream Hollywood studio will be very different from a UK action/adventure movie.
➜ A newpaper article about the Labour party annual convention will be very different in *The Times* from in the *Guardian*.
➜ A government-sponsored website about the dangers of chatrooms is likely to be very different from a site giving information about how to get access to good chatrooms.

Check the net

The *Guardian* newspaper has maintained a detailed archive of materials about the Iraq war which has some interesting articles about the reporting of the war. You can access the archive at http://media.guardian.co.uk/iraqandthemedia – a report about this CNN and Reuters story can be found in this archive.

Links

There is a lot more information about institutions in the key concepts section on pp. 14–15.

Checkpoint 3

Sometimes TV programmes move from BBC to ITV or the other way around. They usually have to make some changes to fit the new context. Can you think of some recent examples?

Exam question answer: page 49

Think of a typical British film, such as *Four Weddings and a Funeral* or *Lock, Stock and Two Smoking Barrels*. In what ways is it typically British? (1 hour)

Answers
Media language

Introduction to media language

Checkpoints

1 The front cover of a book is an important tool in selling that book to an audience and clearly establishing the ideology and genre of the book. A textbook has to appeal to both students and teachers and clearly establish itself as authoritative but fun and engaging for the subject. This is usually done with a combination of words and images.

2 Most soap operas dress their characters very carefully. Both *Hollyoaks* and *Neighbours* have a range of characters and will costume them to suit particular roles or stereotypes – older characters tend to be eccentric or very bland while the younger characters are usually much less extreme. 'Bad' characters usually dress distinctively as well.

Exam question

A textual analysis of this trailer would logically start with a focus on context of production and consumption. By starting with an identification of the target audience and then relating this to the institution, it becomes easier to justify why it was constructed in a particular way, not just how it was constructed. Since the trailer was made for a particular cinema-going audience, it uses generic codes and conventions to attract that audience. In particular, the Bond films have an iconography and narrative formula which is consistent across all the films and an essential part of the Bond subgenre. Therefore the function of the teaser trailer is to engage the audience, reassure them that their expectations will be fulfilled and provide enough enigma about exactly how Bond will resolve the crisis this time to hook them into wanting to watch the film. What is significant about this trailer is the balance between creating this enigma and showing enough of the expected conventions to appeal to the audience.

It is sometimes difficult with film material to relate the media language and representations used directly to the target audience because the target audiences can be very broad, so an approach through genre is usually more focused. By constructing an analysis around this duality of function (enigma and formula) it is easier to explore the key concepts of institution, representation and ideology through the media language employed and consider how this will appeal to fans of the subgenre.

3 You should be able to identify very quickly what sort of reading you made of different television programmes you watched yesterday. It is likely that you received a preferred reading from programmes you chose to watch and an oppositional reading from programmes you had to watch because other family members were watching them, for example. If you were watching a particular programme with other members of your family, you may well have all read it differently because you are different ages and have different cultural expectations.

Exam question

Men's lifestyle magazines have only become popular fairly recently. This niche market has become a very important part of the magazine marketplace. The magazines are image-heavy, with high production values, and cover a wide range of content – unlike hobbyist magazines, for example, which tend to be more content-led and less weighted towards production values.

The key to analysing one of these magazines is to deconstruct it through audience identification. In order to find and maintain a niche within the marketplace, each magazine has to construct a particular ideology and image which in turn constructs a particular ideal reader for the audience to aspire to. The primary ideology of all lifestyle magazines is aspirational – which is why the content can be so wide-ranging. The audience is likely to be defined quite closely in terms of age, family background, job role, socio-economic status and psychographic profile. By identifying the niche market in this way, the magazine publishers can construct an aspirational ideology which appeals to that audience directly and thus secure their market position. So, for example, the type of cars or holiday reviewed in the magazine would depend on the niche market, as would the weighting that might be given to issues such as seeking promotion at work or articles about female celebrities (or indeed, the choice of celebrities).

Stronger candidates will be able to incorporate all this into their textual analysis, and to justify points made in terms of why the magazine was constructed this way and how the ideology is reflected in this construction. Less secure candidates will provide a reading of the text but without following the reading through to explore the motivation behind this construction and the way that institutions construct target markets by these aspirational devices.

Audience

Checkpoints

1 The BBC has four different mainstream radio stations because in this way they are able to reach four very different audiences. The audiences tend to be categorised by age and socio-economic status as much as by identity: Radio 1 being aimed at the youngest audience; Radio 2 at an older audience; Radio 3 at a much more sophisticated and affluent audience and Radio 4 at older professionals.

2 'Mediate' means 'adapt' in this context. An institution will establish a particular meaning and ideology for a media text but this meaning will be mediated by the audience because they will receive it within their own cultural context and this will shape their response.

Representation

Checkpoints

1 Representations of students in school brochures usually show happy, smiling students wearing very neat uniform and looking very interested in their lessons – just the image a parent wants to see!

2 From your analysis you will see that daytime TV is aimed at housewives; early afternoon on the mainstream channels is mostly aimed at children; early evening programmes are aimed at people who have just got in from work and the major entertainment programmes are usually shown after 9 p.m. to catch the biggest audiences.

3 New York is represented as much more thoughtful and reflective since 9/11. Previously, New York was

represented as quite brash and New Yorkers as quite assertive and loud.

Exam question

Lifestyle documentaries are almost always aspirational (as with lifestyle magazines) and so seek to create ideologies of lifestyles that viewers can aspire to. A range of representations is usually used to communicate these ideologies. For example, if the programme is about people finding their ideal house, there will usually be a range of shots of their current house, with an emphasis on the negative aspects – for example, mise-en-scène which makes the house look small and cluttered. Montage, camera angles, editing and commentary can all be used to represent the current lifestyle as less than ideal as represented by place and therefore prepare the viewer for this to be resolved by finding a new property. Simple devices, such as filming a house in sunshine or in rain can create a positive or negative representation.

In aspirational programmes of this kind, several houses are often displayed to choose from and so the editing is often used carefully to represent each house differently and usually very positively. If a house is run-down but the area around the front door is attractive, with plants, for example, the use of a tight camera frame and smooth editing between close-ups and wider angles which only show the doorframe area can still establish a very positive representation. This might be very different if it also included sight of broken guttering, peeling paint and missing roof tiles. Such a representation would be less than aspirational!

So, stronger candidates here would identify what ideologies are being constructed and consider how the representations are used to create and sustain those ideologies in the text. You should be able to deconstruct the representations of people and place being established, suggest why they have been represented this way and relate that to the institution which has created the text and the ideologies they seek to convey – within the context of how it relates to the target audience.

Semiotics

Checkpoints

1 The colours used on the front cover and as a primary theme throughout an edition of a women's magazine are always seasonally related. So winter editions will use deep reds and golds against whites for example, spring editions use pastel shades and lots of yellow, autumn editions use shades of green, russet and red for example, reminding us of leaves falling off the trees and summer editions use bright, summery colours. By linking the cover to the season, the publishers can emphasise how contemporary the magazine is and why the audience should buy it.

2 An image of a gun with a red covering of some kind would probably lead you to expect a gangster film or possibly a horror film (although horror is more commonly associated with knives).

3 Indexical signs might include the names of particular programmes; iconic signs might be the use of an image of a lead character as a link to further information about them. A symbolic sign might be a sign representing 'back

to the home page' or a pencil shape to indicate 'click here to send us an email'.

Exam question

To provide a semiotic reading of a TV listings magazine, it makes sense first to select what material you will deconstruct in this way. A fair range of material to look at would probably be the front cover, contents page and a single feature article (across two pages maybe) to give a range of signs and contexts.

Remember that you don't need to identify each and every sign on each of these pages and that there is no perfect analysis of types of sign.

You might start by making some notes and seeing if you can start to identify some signs under each of the four headings. You may find that you are not sure which heading to use for some signs – some may even be multisemic and have a range of meanings behind them! Once you have started to think about what signs are being used, you can think about writing your answer, looking at why these signs are used. For example, is there a fair range of signs (suggesting perhaps a broad target audience) or are there a lot of symbolic or arbitary signs, which might suggest a niche market with a lot of shared cultural capital? What level of signs is related to the institution? Remember to justify your comments in terms of target audience.

When you structure your answer, it may be helpful to start by identifying the meaning which is constructed and move on from there to explore how the sign system establishes this meaning. The advantage of this approach is that you can structure your answer in a more flexible way, rather than ending up writing a list of signs and not being able to develop a coherent argument.

Narrative and genre

Checkpoints

1 Most adverts tell a story – it might be: 'This product will revolutionise your love life/family, etc' or the enigma: 'You need to find out more about this', but a clear narrative or story context is a very important part of creating an image for the product.

2 Story time and discourse time will often differ at the beginning of a typical Hollywood film because the film is trying to establish a complete context for the narrative, so uses both to create the narrative but also establish ideology, etc.

3 It is usually very simple to identify the genre of television programmes from their listing – this is often intentional as audiences will usually watch a lot of programmes in a particular genre which appeals to them.

Exam question

An analysis of the narrative being established by any magazine front cover is likely to start by identifying enigma codes being created to tempt the reader into buying and reading the magazine. These are usually the short teaser tags down the sides of the front cover, identifying stories within the magazine.

Answers should also explore the use of multi-strand narrative (lots of different articles) to create the wide range of niche-related topics within the magazine.

The readers of this magazine also expect certain narrative codes such as conventional discourse, e.g. the way readers are described as 'ladies' and the emphasis on the historical context of production for the magazine.

It is also fairly straightforward to analyse a magazine front cover in terms of narrative theory – for example, the use of binary oppositions such as those used in tag lines on a front cover or the way that a headline about a reader's experiences may pose a question which will be resolved by the narrative below.

The focus in this answer is to identify the ways in which meaning is constructed in terms of telling a story – not just through visual representations, ideology and media language. The narrative of the magazine is very important to the reader – unless they value the content, they will not value the media language, so the institution must make sure to create the narrative in the right way.

Lights, camera, action!

Checkpoints

1 The mise-en-scène at the beginning of the film establishes the simple but safe life which is about to be disturbed, the role of John Wayne as the protagonist and the isolation of the Indian character very quickly. This is done by editing, camera angles and framing in particular.
2 The drum beat at the end of each episode of *Eastenders* signifies the tension and drama of the cliffhanger. The actors in the programme are always keen to know who has the drumroll at the end of each episode!

Exam question

The title sequence is always a good starting point to study any text in detail because it is the 'shop window' for the text, which defines the context and ideology, sets up audience expectations and defines the representations and codes and conventions which the audience can expect in the programme. A title sequence for a news programme has to attract the audience's interest, create tension, establish the ideology of the programme, suggest the representations which will be used and the media language to be expected – and construct a perception of the institution for the audience.

An early evening news bulletin for Sky or for BBC One will start from a very different institutional context and this must be established by the title sequence, and therefore communicate ideologies to the audience, e.g. about the balance of hard and soft news in the reports or the emphasis on entertainment which might be expected. The representations to be expected in the programme can also be identified in the title sequence. If there are a lot of international images or images of policiticans or political buildings, this would suggest a certain ideology about what slant on the news might be given by the programme. The soundtrack is likely to be non-diegetic for a title sequence, although some bulletins use cutaways to a presenter for headline stories as part of the title sequence to create tension in the audience and establish gravitas for the programme.

Moving image

Checkpoints

1 For each stage in the OS (opening sequence) of the news broadcast, you can look at the mise-en-scène, camera angles, framing and editing to construct meaning and context for the broadcast. A tremendous amount of information – in particular in relation to ideology – is communicated visually.
2 Representations in a news bulletin tend to be of 'goodies' and 'baddies' – apart from the presenter, who has to be represented as formal and formally dressed to impart the required gravitas to the reporting.
3 The rhythm of the action is often disrupted significantly at the beginning of a horror film, partly to unsettle the audience and prepare them for shocks to come but also because this hooks the audience into accepting the narrative as it stands (too disruptive to reject these climax points) and therefore ensure a more positive audience.

Exam question

In many ways, this is exactly the same question as has been asked about print texts in that it is primarily asking about how visual signs create meaning for the audience, establish the genre codes and conventions and therefore set up audience expectations for the film.

When looking at moving image texts, stronger answers will always remember that, while it is important to look at basic visual signs such as use of colour or proxemics, it is also important to remember how the action is structured – in other words, how movement and editing construct meaning. For instance, a series of high- and low-angle shots during an interplay between two characters can quickly establish for the audience where the power in the relationship lies. In this sequence, a variety of methods is used to suggest the power of Neo and his status in relation to Mr Black.

Familiarity of soundtrack and costume, for example (because this is a sequel) can all satisfy audience expectations since they reassure the audience that the genre will not be subverted but sustained. Similar editing for action sequences, similar credit tracks and montage all add to this reassurance and set up the disequilibrium that must be resolved by the end of the film.

Print texts

Checkpoints

1 The masthead on any magazine is an important signal of the ideologies in the magazine. The choice of colour and font will reveal a lot about the ideology, more than the content of the magazine. If the masthead also includes information such as the price, this reveals even more about the ideologies promoted.

2 If this were a spring photo, for example, it would make sense to have a pastel suit to reflect spring colours and a good starting point would be green, since that colour symbolises rebirth.

Exam question

This is a typical textual analysis question which you might be asked in a final exam paper because it is much more general. You should now be able to see that it is simply asking the same questions as you have already been asked in relation to specific key concepts, but here you are assumed to be covering all of them.

It is not possible in an exam situation to cover everything in your answer, so planning is vital. You need to start by making some brief notes for yourself about the target audience for each text (and any possible overlap) and the institutional context of each text (i.e. who publishes it, price, etc). With magazines, it is often important to note when they were published – as we have noted before, the colour scheme, etc is likely to reflect seasonal issues to make the magazine seem very contemporary and attractive. Once you have made these basic notes you can probably identify the ideology being constructed by each magazine and therefore the representations being used to establish that ideology within this institutional context. By doing all this and giving detailed evidence of the media language being used to communicate to the audience you will be providing a detailed reading of the sort examiners are seeking.

Deconstructing new media

Checkpoints

1 The website for a new film is likely to reuse the same graphics as other promotional material and to make use of screenshots, etc as downloadable resources. It is important for the website to be immediately associated with the film but it will probably forge its own identity in relation to perceived audience expectations.

2 There are many different genres of computer game, such as simulation or action/adventure (as per films). The audience is progressively more male as the games get more violent. Female players end up getting sidelined anyway!

Exam question

Deconstructing a new media text is done in the same way as for a conventional text – the closest parallel is to deconstructing a print text. You can structure your analysis in a similar way in terms of analysing institutional codes and conventions and ideologies in relation to the target audience, but you need to extend your analysis to consider layout a little more fully, also navigational structure, etc. For example, if you are looking at a webpage, it is important to remember that viewers may look at the same webpage with different-sized monitors or different web browsers, so pages need to be constructed to support this. Is it easy to access the information that you require, or is it difficult? Is scrolling likely to be acceptable to the target audience for this text?

You can also look at how navigation is structured – are links mainly images or text? How does this relate to the target audience? How does it reflect the ideology of the site?

What representations are used on the pages? How are moving images (e.g. flash animations or scrolling text bars) used to convey meaning to the audience? Do they complement the pages or are they a distraction?

If you are deconstructing the OS or trailer for a computer game you will probably find that you deconstruct it as a moving image text – quite often the OS is not interactive and is presented almost as an animation to establish the narrative and ideologies of the game for the audience. In this case, it makes sense to deconstruct the text in the same way as a moving image text but remembering that the context of construction is to establish audience expectations for an interactive game, for example, not a linear moving image text.

Texts in context

Checkpoints

1 We all receive texts differently in a group situation. Our views are partly shaped by those around us and this can affect our personal readings of a text very significantly.

2 Most reality TV programmes are keen to encourage the water-cooler effect, knowing that people will want to talk about the latest eviction, etc. This is good publicity since other people will then become hooked into watching the series, expanding the original audience.

Exam question

As we move beyond simple textual analysis, it becomes important to be aware that texts are created within a context, not in isolation. This will influence and shape the text and should be considered when analysing it. In this question, you are being asked to identify which typical codes and conventions of British film you can identify in the text. What ideologies and representations are used? You need to think about obvious iconography such as location and speech but also to think about narrative devices, non-realist conventions (such as characters speaking direct to camera) and narratives which do not always fully resolve. There can often be more political ideologies in a British film and characterisation may be more complex than in equivalent Hollywood films.

There are particular genres of British film, such as British Asian films or social commentary films, and you will need to place a British film within the relevant genre and identify the use of generic codes and conventions in relation to the target audience if you are to be able to explore the text in any depth.

The central focus of the answer therefore must remain on the ways that the British context shapes the text and all your deconstruction should relate to that. This involves much more careful analysis than simply deconstructing the text in isolation.

Revision checklist
Media language

1	Understand what is meant by target audience	Confident	Not confident. **Revise** pages 28–29
2	Understand how people can be represented	Confident	Not confident. **Revise** pages 30–31
3	Understand about types of sign used in a sign system	Confident	Not confident. **Revise** pages 32–33
4	Understand what is meant by the genre of a media text	Confident	Not confident. **Revise** pages 34–35
5	Know how to deconstruct mise-en-scène and sound	Confident	Not confident. **Revise** pages 36–37
6	Understand how editing creates meaning	Confident	Not confident. **Revise** pages 38–39
7	Know how to deconstruct a print text	Confident	Not confident. **Revise** pages 40–41
8	Know how to deconstruct new media	Confident	Not confident. **Revise** pages 42–43
9	Understand contexts of production and consumption	Confident	Not confident. **Revise** pages 44–45

Media forms

Each media form that you are likely to meet during your course has its own elements of media language, technical codes and conventions, typical genres and audiences and common ideologies and you will need to be able to write about these at AS and A2 level. In this section, we will look at some of the most common media forms in relation to the key concepts and their form and function. This is clearly not a definitive list and, as with all areas of media studies, you will notice overlap with other sections, but this should give you some good starting points about the differing media forms which you need to be able to work with.

As you work through this section, try to be aware of some of the media terminology which is used on different pages. You need a secure grasp of media technology to demonstrate your understanding when writing your answers. You will also find that a good knowledge of media terminology will often make it quicker and easier to explain a concept in detail. You don't need to explain what terms mean when you use them – your examiner should know what the words mean!

Exam themes

→ Types of media text

→ Industry sectors

→ Traditional and interactive media

Topic checklist

○ AS ● A2	OCR	AQA	WJEC
Commercial television	○●	○●	○●
Public service broadcasting (PSB)	○●	○●	○●
Reality TV	○●	○●	○●
Situation comedy	○●	○●	○●
Soap operas	○●	○●	○●
Independent cinema	●	○●	○●
Mainstream cinema	○●	○●	○●
Documentary	○●	○●	○●
Advertising and marketing	●	○●	○●
Magazines	○●	○●	○●
Radio	○●	○●	○●
Newspapers	○●	○●	○●
News broadcasting	○●	○●	○●
Video/computer games	●	○●	○●

Commercial television

Whereas once there was only the BBC, now there is an immense range of commercial television available to us. It is important that you know what commercial channels and institutions are involved and some of the differences between them. You also need some understanding of the differences between terrestrial, satellite and cable providers.

Commercial television ●●●

Commercial television services include all the ITV services, GMTV, Channel 4, Channel 5, Teletext and local services on terrestrial television, as well as a wide range of cable, satellite and digital services. Although the regional services were originally all independent, most are now owned by either Carlton or Granada.

The Broadcasting Act 1990 established the Independent Television Commission to regulate commercial television and the Broadcasting Act 1996 set out its duties and responsibilities in licensing and monitoring analogue and digital services. The ITC ceased to exist in 2003 and was replaced by OFCOM as a regulatory body for all communications organisations in the UK.

Although Channels 3, 4 and 5 and the satellite and cable channels are funded by advertising revenue, they are still accountable and must provide a range of programmes for different social and cultural groups. **Channel 4**, in particular, has a particular remit about the types of programme which it must broadcast. Of course it can make decisions about what type of programmes to show at what times. If it shows popular programmes during peak times it can make more money from the advertisers.

For example, one of the most expensive slots on ITV is the advert breaks during a major football match, or during a major sporting event. Because the advertisers know that the audiences for these events will be huge they are prepared to pay a lot more money to be able to reach that audience at that time.

There are ever-increasing numbers of satellite and cable channels. This has led to the emergence of **third-generation** cable and satellite channels which are far more **narrowcast**, with particular themed output.

As commercial broadcasters seek to fill these niche markets, they have started to re-show previous PSB broadcasting (think of the programmes shown on UK Gold or Turner Classic Movie, for example). However, in general, arts, education, multi-cultural programmes, investigative documentaries, current affairs programmes, natural history programmes and the like continue to be less common on commercial channels because they are less profitable for advertisers. There is instead more frequent use of non-UK programmes to fill these channels – in particular, US programmes and also there is increasing repetition of old programmes to fill the airspace.

Cable and satellite television

In digital homes viewers are now able to choose from many channels serving niche interests and are increasingly acting as their own schedulers by using new technologies such as **TIVo** to control which programmes they view. Again, this has implications for advertising revenues.

Commercial institutions are good at generating very popular programmes, including dramas and serials, but there is little variety – hence the large number of game shows, reality TV shows, etc. Serials tend to be police and hospital thrillers with less variety than on the BBC and more use of **star vehicles**.

Commercial channels are generally interested in programmes with maximum audiences. In recent years ITV has been prepared to compete against the BBC in various ways, for example, by costume dramas – although, perhaps inevitably, they have also made more popular programmes such as *Big Brother* and *I'm a Celebrity*. In return, the BBC has created more popular programmes, leading to complaints that it is not fulfilling its PSB Charter.

This has led to accusations of **dumbing down** as broadcasters concentrate on programmes which generate the maximum audiences so that they can get maximum advertising revenue. Whereas once channels would not go 'head to head' with popular programmes, now they will deliberately schedule against each other. Of course, one reason for this is that almost all households have a VCR, so are happy to record one of the programmes, which creates different expectations. As recording programmes becomes more flexible, with systems such as TIVo, this may change again.

Bookshop of the air

There is an argument that the number of new channels now available to digital viewers will lead to digital television offering a '**bookshop of the air**', giving the viewer so many choices of programme that **free to view** channels, including PSB channels, will become redundant as viewers become their own schedulers. They would be able to choose from hundreds of channels, providing high-quality programmes for all community groups.

The problem is that this would be expensive to provide. The 'bookshop' is an illusion, because it is neither commercially attractive, nor does it fit with the public's habits or expectations. **Free to view** channels, including PSB channels, are unlikely to disappear in the near future.

Checkpoint 1

What is a star vehicle?

Checkpoint 2

What do we mean by 'dumbing down'?

Exam question   answer: page 80

Compare the TV schedules across a range of free to view and satellite channels for one day. Do you think there is enough evidence to argue that television is dumbing down?   (1 hour)

Public Service Broadcasting (PSB)

In this country, there are two forms of terrestrial television and a range of cable and satellite channels. It is important that you know the difference between the different organisations involved in producing television and how this affects their output.

Principles and requirements of PSB

PSB or Public Service Broadcasting is based on of eight basic principles:

→ It aspires to universal availability – i.e. to reach the whole nation.
→ It aspires to universal appeal – i.e. to make good quality programmes that appeal to some of the people all of the time, and all of the people some of the time.
→ It recognises citizens, not merely consumers, and that they have responsibilities as well as rights.
→ It recognises and seeks to cater for substantial minorities as well as majorities in the television audience.
→ It seeks to inform and to educate as well as to entertain, not merely to entertain.
→ It should be independent, particularly of political and commercial interests.
→ It should encourage competition in good programming rather than competition for numbers.
→ It should liberate rather than restrict the programme maker.

The overall requirement for PSB broadcasting, not just in the UK, is that it should 'inform, entertain and educate' in return for receiving finance through the licence fee. PSB is about providing a public service, which means:

→ a service for the public (i.e. available to the public)
→ a service in the interests of the public (i.e. with content which is for the public good)
→ broadcasting which is appealing to members of the public (i.e. individual people). In other words, it must be interesting as well as good for us!

Along with this, it has to appeal to all groups in society, providing content the audience want as well as need, across all social, cultural, ethnic and other groups in our society equally.

In this country, the licence fee that has to be paid for owning or renting a television set is used to fund the BBC. However, there is also a PSB expectation for Channel 4, which has a requirement to provide a range of programmes for minority groups in society. ITV, S4C (Wales) and Channel 5 are also required to abide by these terms. All the 'commercial' channels, which do not receive licence fee funding, have to generate their income in other ways.

Checkpoint 1

What is the television licence fee?

Check the net

You can find out all about the BBC's charter at http://www.bbc.co.uk/info/policies/charter/

Check the net

You can find out more about S4C at http://www.bbc.co.uk/wales/

The BBC

The British Broadcasting Company, as the BBC was originally called, was formed in October 1922 The first general manager was John Reith. He envisaged an independent British broadcaster able to **educate**, **inform** and **entertain** the whole nation, free from commercial pressure. In 1927, the British Broadcasting Company became the British Broadcasting Corporation when it was granted its first Royal Charter. The BBC was a radio institution to begin with and did not start to broadcast television programmes until November 1936.

Although the BBC broadcast a wide range of programmes at this time, the newspaper industry stopped the BBC from broadcasting the news. Bulletins were prepared by the news agencies, and could only be broadcast after 7 p.m. – so as not to upset newspaper sales.

On June 2 1953 a single event changed the course of television history. An estimated 22 million TV viewers watched the coronation of Queen Elizabeth II.

BBC Two began broadcasting in black and white in 1964 and in colour in 1967. Colour was introduced for BBC One in 1969. In 1972 the CEEFAX text service was introduced and subtitling began in 1979. 1979 also saw the introduction of BBC Worldwide, the commercial arm of the BBC, marketing videos, books, audio and magazines.

Recent developments have allowed producers to choose between BBC suppliers and the outside market for their facilities. Digital expansion has included the launch of BBC Four, CBeebies, CBBC and BBC Three. Interactive television started with coverage of *Wimbledon 2001* and continues to grow, now becoming commonplace in documentaries, for example, after the success of *Walking with Dinosaurs*.

The most obvious difference between the BBC and commercial channels is that there is no paid-for advertising on the BBC. Profits to fund the new channels and high-concept programmes such as *Walking with Cavemen* are now generated not only by the licence fee but by BBC Worldwide, which sells BBC programmes globally.

The BBC also produces a range of magazines, educational materials, and so on, which are all produced commercially, outside of PSB restrictions and earn substantial amounts of money. However, government approval is required under the Charter for some of the BBC's commercial activities. No public funds can be used for commercial activities but profit can be made from successes, like the Teletubbies, and re-invested in programme making. The result is that more money can be spent on publicly funded services without necessarily having to increase the licence fee.

The jargon

Interactive televison is available through your satellite or cable connection and means that you can watch several different things at once, perhaps watching three different sports during the Olympics using three different windows on your screen as well as being able to vote for contestants in reality gameshows, for example.

Checkpoint 2

Does the BBC offer any subscription television channels?

Exam question   answer: page 80

Do you think that the licence fee is still a legitimate form of funding for the BBC?   (1 hour)

Reality TV

Reality TV is probably one of the most significant new genres which have developed within television. It has grown out of the documentary tradition of cinema verité but has now become a genre in its own right, popular with television producers and audiences alike. In this section we look at the development of the genre and the codes and conventions now associated with it, as well as some of the most significant programmes in the development of the genre.

Checkpoint 1

What is cinema verité?

History of reality TV ●●●

Reality TV developed from early 'real life' television such as *Candid Camera* which started in the US in 1948. Probably the first show which would be regarded as being within this genre was *Wanted*, which ran in 1955 and 1956. Host Walter McGrew outlined the crimes of fugitives and interviewed their relatives and law enforcement officers working on the cases. The best-known 'surveillance' example of the early genre was probably *An American Family*, in 1973. The Loud family were filmed for seven months, totalling 300 hours of footage. Only 12 of those hours made it to television (demonstrating that significant editorial decision making has always been part of the process). During the run of the show, the marital breakup of Bill and Pat Loud and the coming-out as homesexual of their son Lance was shown. The family complained bitterly that they had been misrepresented on screen.

In 1974 *The Family* was made in the UK, following the working-class Wilkins family of Reading. One of the best known global programmes within this new genre was the Australian *Sylvania Waters* in 1992, about the *nouveau riche* Baker-Donaher family of Sydney. As with the first US programme, both families complained that they had been misrepresented and false perceptions constructed by the editors.

A later development of this type of reality TV series, following a family, has been the **celebrity lifestyle** programme – probably one of the best-known examples is *The Osbournes*. Celebrity reality game shows are also very popular.

Types of reality TV programme ●●●

Docusoaps

Links

Look at the documentary section on p. 67 for more information about docusoaps.

In a docusoap, the viewer (and therefore the camera) are passive, following people in their daily lives at home and/or at work. Early successful examples of this were *Airport* and *Paddington Green*. These docusoaps overlap significantly with documentary, also using codes and conventions from soap operas – hence the name.

Voyeuristic TV

The second type of reality TV programmes are the voyeuristic programmes where the participants are not aware that they are being filmed or set up in any way. Obviously, *Candid Camera* is the earliest example of this genre but there are many versions still being made today – especially on cable channels. The viewer is the 'voyeur', watching but not able to contribute or affect events in any way.

Reality game shows

Reality game shows are probably now the best known examples of reality TV. From the success of the first *Big Brother*, there have been many reality game shows of different kinds, such as *Castaway*, *Temptation Island* and *Survivor* as well as obvious competitions using a reality format such as *Fame Academy*.

These programmes incorporate voyeuristic elements (hidden cameras *et al*) but also game show structures such as contestants being voted off, either by their peers or by the audience. Of course, they are not really voyeuristic in that all the contestants will have volunteered to be in the programme; material is always heavily edited before being shown and the cameras are less than hidden in many cases. This may actually mean that the most popular form of reality TV is more heavily constructed than many programmes which do not pretend to be real.

Celebrity reality game shows

As in the original 'reality documentaries', celebrities quickly became the primary contestants for reality game shows constructed to attract large audiences. *Celebrity Big Brother* was a bigger hit than the original programme and the several series of *I'm a Celebrity, Get Me Out of Here* continue to gain increased audiences. The combination of voyeuristic footage, celebrities competing to complete various tasks, individual interviews with the contestants and a clear structure and context provided by the presenters has created a very successful formula.

Codes and conventions of reality gameshows

➜ Contestants are voted off one at a time to leave a winner.
➜ There are private interviews with contestants every day, allowing them to express their private thoughts as if in an interview to a tabloid newspaper.
➜ Contestants have to complete a range of tasks – often to 'earn' food for that day.
➜ Although contestants are filmed 24 hours a day, programmes are edited down to maybe just half an hour of material, broken up with interviews and slots from presenters to make a one-hour programme – the identity of the contestants is therefore constructed by editors.
➜ Viewer participation is increasing – viewers can vote off contestants by phone, by text message, by email or on interactive TV. This audience participation is helping to grow the genre.
➜ Contestants become C list celebrities simply by taking part in the programmes. Most gain little in the way of prizes or prize money so it is the fame that attracts them to the game.

Action point

What do you think motivates celebrities to appear on a reality game show?

Checkpoint 2

What do we mean by a C list celebrity?

Exam question   answer: pages 80–81

Watch an episode of a contemporary reality game show. In what ways do the editors construct contestants into particular character roles? Is this helpful for the audience or does it manipulate audience responses too far?   (1 hour)

Situation comedy

Situation comedy (sitcom) started on radio – as did most comedy genres. It is an important genre to study because it is very dependent on stereotypes and farcical situations to create comic effect – which also makes it good to study while you are learning textual analysis. There are different sorts of situation comedy but the basic structure remains the same across all sitcoms.

Key features of sitcoms ●●●

A **sitcom** or **situation comedy** is a television (or radio) comedy with certain key features:

→ The same characters appear in every episode and usually do not change or develop much.
→ Episodes are usually for a 30-minute slot.
→ The set for the sitcom is usually the same in each episode.
→ Sitcoms usually centre on a family home or workplace.
→ Characters in sitcoms are usually very stereotypical.
→ Some British sitcoms are situation-led, so various mishaps happen to the main character in each episode.
→ US sitcoms are often character-led.
→ There is generally a moral conclusion at the end of each episode of a US sitcom.
→ Most characters in sitcoms have clichéd sayings which they will say at least once in every episode.
→ Although there are ongoing storylines, the primary plotline is resolved in each episode.
→ There is a main problem to be resolved by the end of each episode.
→ Ideologies of US sitcoms are far more idealistic and aspirational than those of UK sitcoms.

Sitcom history ●●●

Sitcoms started on the radio in the 1920s. One of the most successful early sitcoms was *I Love Lucy*, starring Lucille Ball, which is still shown on cable channels today. It was the first sitcom to be recorded, not just broadcast live. One of the most striking things about the sitcom was the very outspoken female lead. This was a big contrast to previous programmes.

I Love Lucy made it to the UK in the 1950s and introduced much more sophisticated camera work than had been seen in the UK before.

The most successful of the early 'homegrown' sitcoms was *Hancock's Half Hour* which had moved from radio to TV in 1956.

Hancock's Half Hour introduced the British working-class sitcom which centred on simple, everyday plot events, class issues and less middle-class sets than used in the previous version of US sitcoms. The aspirations of characters in these British sitcoms were very different from those in the US versions – *Steptoe and Son*, for example, centred on two less than attractive, workshy men in a very ugly environment – nothing like the clean-cut, US sitcoms with a moral lesson at the end of each episode.

Action point

Pick an episode of a contemporary sitcom and analyse the codes and conventions of the sitcom. Is there a laughter track, for example?

Checkpoint 1

Are sitcoms recorded live?

Action point

It is fairly easy to get hold of copies of early sitcoms from the library or from video suppliers and you should try to watch at least one so that you can see how little the codes and conventions have changed.

1970s UK sitcoms included *Porridge* (set in a prison), *Are You Being Served?* (set in a failing department store) and *Dad's Army* (set in the fictitious town of Warmington on Sea during the Second World War). Again, unlike US sitcoms, there was no sense of drawing conclusions or learning from mistakes at the end of each episode. Events simply continued each week as before.

In the second generation of British sitcoms, such as *Red Dwarf*, *Blackadder* and *The Young Ones*, the sitcom changed somewhat from the very closed and insular settings featured in the earlier, family-led programmes to sitcoms which challenged the audience and were often driven by the unexpected responses characters might make to a situation. It became the responses of the central character(s) to events around them that generated the comedy rather than the situations themselves – think of characters such as Basil Fawlty, Blackadder, Edina and Patsy and DelBoy, for example. *The Simpsons* uses many elements of sitcom in the cartoons – a postmodern take on the genre and characters.

What is the appeal of sitcoms?

We watch sitcoms to make us laugh. We laugh at the things which happen to the characters, whether the comedy is generated through the situation, through the relationships or through the eccentric behaviour of the main character.

→ In some conventional sitcoms such as *My Family*, it is the endless problems of family life which we laugh at.
→ In a character-led sitcom such as *Father Ted* we laugh at the stereotypes presented to us and much of our humour comes from the non-stereotypical situation.
→ In *Absolutely Fabulous*, the comedy comes through the bizarre and grotesque characters of Edina and Patsy.

Types of British/US sitcom

→ The most conventional type of sitcom is the family situation such as *Malcolm in the Middle* or *2.4 Kids*.
→ Another typical type of sitcom is the seemingly normal family with some bizarre secret or unusual behaviour, such as *My Hero* or *Bewitched*.
→ Many UK sitcoms are class-based, such as *Only Fools and Horses* or *Keeping Up Appearances*.
→ US sitcoms such as *Friends* borrow elements from soap operas to keep the genre fresh, such as cliffhanger endings to episodes.
→ Some UK sitcoms are based around a central figure who is unusual or eccentric in some way and often has strange ways of behaving (Basil Fawlty) or particular catch phrases such as 'I don't believe it!' in *One Foot in the Grave*.

Checkpoint 2

Why is *The Simpsons* sometimes described as a postmodern sitcom?

Action point

Try to find images of these well-known sitcom characters. How important is the visual representation of these characters?

Exam question   answer: page 81

Watch an episode of a contemporary British sitcom. In what ways does it conform to or challenge its genre?   (1 hour)

Soap operas

In this section we will look at soap opera. We will look at the history of soap operas and their central position in the television schedules and consider their codes and conventions in relation to the key concepts.

History of soap operas ●●●

Soap operas started as serialised dramas on radio in the US in the 1930s. The name 'soap opera' was given by the newspapers to these dramas because they were often sponsored by companies which made washing powder, soap and cleaning materials. The genre quickly expanded globally, not just in the US.

Codes and conventions of soap operas ●●●

→ They have a serial narrative told through a series of episodes.
→ The narrative progresses between episodes.
→ There is a cliffhanger at the end of each episode, to hook the audience into watching again.
→ Characters change and develop over time in a soap opera.
→ Unlike in a serial, there is no end to a soap opera (although see below).
→ The audience usually identifies with particular characters and follows their progression over time.

Open and closed soap operas ●●●

Open soap operas continue with no apparent end point – think of *Eastenders*, for example, which continues to be shown several times each week, without any end date. Closed soap operas, however, do move to a final conclusion, although this may be a long drawn-out progression. A closed soap opera of this kind is less common on US/UK TV but is very popular in Latin America and is known as the *telenovela*, which suggests in the name a conclusion to the narrative. One of the key attractions for a closed soap opera is finding out what happens in the end, just as with a good novel.

Soaps and audiences ●●●

Traditionally, soap operas are watched by women rather than men and the socio-economic status of the audience has tended to be lower. Given this audience and the perception that soap operas offered little to 'inform' or 'educate' an audience as well as 'entertain', they have always been seen as a 'low-culture' form of television.

Yet, the multiplicity of narratives or **metanarrative** of a soap opera, with many narratives interlinking all with different construction, climax points and character functions make it probably one of the most complex genres on television. In fact, the preknowledge of the multi-narrative to date is a central factor in the stability of the audience, who will engage with the soap over many years. The audience is usually also very knowledgeable about narrative history within a particular soap.

To some degree all soap opera metanarratives have key conventions in common:

Action point

Watch an episode of your favourite soap opera and try to identify as many of the codes and conventions as you can. How important are these to your enjoyment?

Checkpoint 1

What is a metanarrative?

- → fantasy rather than everyday life, especially in the supersoaps
- → central focus on family conflicts and romance
- → several plots at different points in the resolution of the disequilibrium. There is often a primary 'shocker' story leading the metanarrative
- → storylines relating to issues such as domestic violence, teenage pregnancy, drug problems, etc
- → providing a range of characters of all ages and backgrounds to allow all audience members to feel they can identify with a particular character 'like me'.

Subgenres of soaps ●●●

American supersoaps

The supersoaps, such as *Dallas* and *Dynasty* were a particular, fantasy-driven, subgenre of soaps with wealthy, scheming families but still the same form of metanarrative and representations as with kitchen sink soaps. For example, *Dallas* had the extended family (the Ewings), complete with patriarch, matriarch, good son, bad son and in-laws – all of whom lived in the same house. The narratives were constructed around these core family members. The producers were seeking to attract the conventional female audience in this way but also sought to attract a male audience with Ewing Oil's boardroom intrigues. *Dallas* was so successful it was the first US serial to be marketed globally and spawned many debates about US cultural imperialism.

UK soaps

The UK soap tradition began in the 1940s with *The Archers* on BBC Radio. At first this soap did 'inform and educate' as well as 'entertain' (fulfilling the PSB charter). The target audience were farmers, who learnt about better agricultural practices through the soap.

Coronation Street, set in a working-class area of Manchester, started in 1960. It is well known for its gritty portrayal of northern working-class life. Indeed, these roots were so strong that all subsequent UK soaps have taken pains to be based in a particular location in the same way – *Eastenders* (started 1985) in 'Walford' in East London and *Brookside* (started 1982) in a fictional close in Liverpool, for example.

Australian soaps

The first Australian soap to be successful in the UK was *Neighbours*, which started over here in 1986. It has elements of both UK and US soaps in its construction – it is based on a neighbourhood in the UK form but is more fantasy-driven than UK soaps – as seen in the weather and the recreational activities shown and the use of 'fantasy' sequences during times of emotional stress.

Example

Dallas was one of the first soaps to deliberately admit the artificiality of the soap opera when they decided after a period of 10 years to bring back a character from the dead – with an episode where his wife woke up in the morning to see him in the shower and said she had just had a dream that he had died . . .

Checkpoint 2

Why are the types of location for UK and US soaps so different?

Don't forget

Australian soaps are sometimes shown at different times of day in Australia (e.g. post-watershed). This can make a lot of difference to the way that storylines are presented and edited.

Exam question   answer: page 81

Watch an episode of a UK soap and analyse it in terms of the use of soap opera codes and conventions during this episode. How is the metanarrative structured, for example?   (1 hour)

Independent cinema

Although most of the films we watch are made as conventional Hollywood movies, there are a significant number of independent films made every year. One reason why you do not see very many of them is because independent film makers cannot always afford the distribution costs. Another reason can be that multiplexes, in particular, do not show independent films because they are owned by major film corporations who do not wish to encourage independent films.

Audiences can also be resistant to non-Hollywood films. In the US and the UK, foreign language films rarely do well because the majority of the audience are not prepared to watch a film with subtitles.

This section examines the main codes and conventions of alternative films and some of the major non-Hollywood film movements.

Checkpoint 1

What does 'non-realist' mean?

Indie films ●●●

The term 'indie' implies that a film will break the rules of traditional narrative and experiment with new or different ways of telling stories to create meaning in a non-realist way. By comparison, Hollywood film is almost exclusively realist in approach.

One of the first film makers to move away from realist film was French director Jean-Luc Godard. He wanted to expose the constructed nature of cinema and challenge his audience in order to make them reflect on their own lives as they watched.

Of course, indie cinema is often adapted by mainstream cinema, and some films with indie characteristics have become huge commercial successes. One of the most famous of these has to be the *Blair Witch Project*, with its use of unknown actors, handheld camerawork, direct address to camera, and lack of resolution.

Instead of the classic Hollywood narrative as suggested by Field, indie films use narrative in different ways:

➔ Action may be explicitly broken into chapters, as in a novel.
➔ Narrative sequence is disjointed in some way.
➔ The narrative is not resolved.
➔ Well-known actors are not used.
➔ Audience cannot empathise with characters.
➔ Close-ups are not used.
➔ Characters move 'outside the frame'.

Action point

Watch one of the classic films from Godard, e.g. *A Bout de Souffle*. In what ways does it use narrative differently?

Non-realist conventions

In mainstream films, the narrative is constructed to reflect reality. In an indie film, the action may be non-realist because there may be multiple layers of narrative or time; flashbacks and dream sequences may be integrated into the timeline or characters may step 'out of the frame' to interact with the audience. Changes to normal time and space (such as flashbacks) are obvious so that they do not confuse the audience. Mainstream cinema tries to make the audience believe in the reality of the events on the screen, avoiding any techniques that would draw attention to the film making process. Indie film seeks to achieve the direct opposite.

European art house

Many critics claim that European cinema's greatest strength lies in producing art house films that offer an alternative to the popular genres (thrillers, action/adventure, sci-fi, romantic comedies, Westerns, etc) produced by the major Hollywood studios. Art house cinema is closely associated with genres and form at odds with the perceived escapism of Hollywood productions. One key difference with European art house cinema is the greater emphasis on the imagination and vision of the film maker, including film makers such as Godard, Truffaut, Fellini, Rossellini, Hertzog and Kieslowski.

Art house cinema is, of course, not purely European and can be found in American cinema too, with directors like Scorsese, Coppela, Tarantino, David Cronenberg and Oliver Stone seeking to produce films more in line with European art house traditions in the auteur style.

British film

There are many genres and traditions within the history of British film. The earliest British films were contemporary with the work of the Lumière brothers and other pioneers and it was British directors who led many of the developments in constructing meaning in documentary and fiction films.

Among the particular traditions, institutions and genres of British film which you could explore are:

→ Ealing comedies
→ Hammer Horror movies
→ *Carry On* films
→ British gangster movies
→ Angry Young Men
→ Social realism
→ Merchant Ivory

Bollywood

Bollywood is gaining significant global audiences, supported by new technologies which allow the easy distribution of Bollywood movies worldwide.

The Bollywood studios began production at around the same time as the US and UK studios, producing a wide range of films, in many different languages. The Bollywood musical, which epitomises Bollywood in the West, is only one part of a tradition which includes surrealist films, social realism and contemporary critiques of Indian society. It was the revival of the musicals filmed in Hindi in the 1990s which brought Bollywood to mainstream attention.

Bollywood is important to study – not least because many directors, such as Luhrmann, have made great use of codes and conventions of Bollywood musicals in their films and the use of colour, sound and light has directly influenced many recent Western films.

Check the net

There is a lot of information available on the web about British film. A good starting point to find out more is www.bfi.org.uk.

Check the net

A good starting point to find out more about Bollywood is the website at www.planetbollywood.com where you will find lots of information.

Checkpoint 2

Which of Luhrmann's films makes most use of Bollywood codes and conventions?

Exam question　　　　　　　　answer: pages 81–82

Watch a contemporary Bollywood musical. What codes and conventions of the genre can you identify? In what ways have these been used in recent Western films?　(1 hour)

Mainstream cinema

Although a complete history and study of cinema is more appropriate for Film Studies than for Media Studies, it can be helpful to have some idea about the history of film, cinema institutions and film genres. If you are studying a particular film genre in depth, you will obviously study it in a lot more detail than can be covered here.

Cinema history ●●●

The Lumière brothers are credited with the birth of cinema in 1895, with the film *Workers Leaving the Lumière Factory*, although the film only recorded factory workers leaving the Lumière factory gate for home or for a lunch break.

The first documentary, *The Life of an American Fireman* (1903), was made by Edwin Porter. It combined re-enacted scenes and documentary footage, and inter-cut between the exterior and interior of a burning house to create tension. His next film, *The Great Train Robbery* (1903) set many milestones – in particular in the way editing was used to create meaning. In a CU at the beginning of the film a baddie shoots his gun directly into the audience, for example. The film also used exterior shots, chases on horseback, a camera pan with the escaping baddies and a camera mounted on a moving train.

Hollywood and the studio system

By the end of the 1920s, there were 20 Hollywood studios, and the silent films were being manufactured, assembly-line style, in Hollywood's 'entertainment factories'.

Even these earliest films were organised into **genres**, with instantly recognisable storylines, settings, costumes and characters to fulfil audience expectations. The emphasis was on swashbucklers, historical extravaganzas and melodramas, although all kinds of films were being produced throughout the decade.

The **studio system** was established in the 1920s (with long-term contracts for stars, lavish production values, and increasingly rigid control of directors and stars). After the First World War America was the leading producer in the world, although the 'factory' system did limit the creativity of many directors. Production was in the hands of the major studios who had consolidated and now controlled all aspects of a film's development – **vertical integration**. By 1929, the **Big Five** were producing more than 90 per cent of global output.

The Big Five had vast studios with elaborate sets for film production. They owned their own production and distribution facilities and distributed their films to their own theatres (or 'movie palaces'). The Big Five were Warner Bros, Paramount, Twentieth Century-Fox, RKO and MGM. Universal, United Artists and Columbia were known as the 'little studios' because they did not have complete vertical integration of all stages of production and distribution. In one form or another these studios still run Hollywood today.

Checkpoint 1

What is a CU?

Checkpoint 2

What is vertical integration?

Contemporary Hollywood

The established Hollywood movie studios (except for Universal and Walt Disney's Buena Vista) no longer directly control production and distribution in the same way, although the development of the multiplex is putting control back into the hands of the large institutions. While studios still dominate film distribution, other areas, including production, filming and financing, are increasingly in the hands of independent studios, producers and/or agents.

The new studio: DreamWorks

The first new Hollywood studio in many decades, **DreamWorks (SKG)**, was formed in October 1994 by Steven Spielberg. DreamWorks has produced many Best Picture winners, including Sam Mendes' *American Beauty* (1999), Ridley Scott's *Gladiator* (2000) and Ron Howard's *A Beautiful Mind* (2001). Dreamworks has a more creative approach to film making than many of the traditional studios, being born from Spielberg's frustration at not being able to make the films he wanted for other studios.

Stars and auteurs

Stars have become increasingly important to the success of a film and the films most dependent on their star actors have become known as **'star vehicles'**. It is now commonplace for the stars of a movie to earn substantially more than the director. They remain popular with the studios because they know that having star names in the movie will ensure audiences and thus revenue.

Along with this has been the development of 'auteurs' (a term first used by Truffaut) – directors who are so influential over the form, meaning and content of a film that they create new meaning or new form for the cinema. The original auteurs, such as Orson Wells or Alfred Hitchcock, clearly deserved the label since they revolutionised cinema in particular ways, but the label is sometimes used now simply as a marketing tool, to attract an audience to see the film.

Independent films

Existing alongside mainstream Hollywood film production are the independents. Most studios have formed independent film divisions (such as Fox's Searchlight division) that make films within the independent tradition, which are artistic, edgy, or centre on serious social issues or themes. These are often made without major Hollywood stars. However, over time, these independents have become more mainstream and institutionalised because of the cultural clash between the desire to make independent films and to make a profit at the box office.

Checkpoint 3

What is a multiplex?

Action point

There are many contemporary directors who are described as 'auteurs'. Looking at the films of one of these directors, decide for yourself whether this description is justified.

"Some writers and some directors are jealous of the stars' glory and the auteur theory is just another attempt to wipe the stars off the screen with words."

Louise Brooks, an early film star, did not agree with auteur theory.

Exam question   answer: page 82

Choose an auteur such as Hitchcock or Tarantino. In what ways can they be described as an auteur?   (1 hour)

Documentary

Documentary is one of the earliest forms of cinema and television and still one of the most important. Early film was originally documentary, such as the Lumière brothers' early films such as *Workers leaving the Lumière factory*. Modern documentary can take many forms but the essential codes and conventions remain the same.

The documentary form ●●●

Although documentary gives the impression of reality, it is important to remember that all documentary is constructed and carefully edited. The skill of the documentary maker comes in the way they can make the documentary seem real and also establish a clear ideology.

Documentary conventions

→ Hand-held camera work, poor lighting (often dark and slightly out of focus) and poor sound (for example, unclear speech which may even need subtitles on screen) suggest that there was no time to set up shots or take care with filming.

→ Editing – disjointed sections or unclear sequencing can imply that the documentary maker has not been able to edit the 'reality'.

→ Use of voiceover or presenter on screen makes it clear that this is not drama.

→ Interviews with participants add integrity and again emphasise that this is factual, not fictional.

→ Use of captions, archive film and photographs/stills create a clear context for the events and again emphasise that this is factual (they almost act as 'proof').

→ Music – this may support the visuals or clash in some way, implying that the documentary maker does not agree with what is being shown or heard.

→ Framing is used to create meaning and establish who the audience should empathise with or disagree with, for example.

Types of documentary

Event documentaries provide a record of a major event such as a music festival or royal occasion. There will usually be a voiceover or presenter and use of montage to maintain audience interest. The London Marathon is usually filmed as a documentary event rather than a sports fixture.

Documentary accounts will follow the events leading up to a particular happening. There may be different accounts or perspectives offered about the chain of events. Historical documentaries are often presented in this way.

The jargon

The voiceover on a documentary programme is often called the 'voice of God' because it frames our responses so fully.

Checkpoint 1

Why is the London Marathon filmed as a documentary?

Documentary journeys may be the account of an investigative reporter following a story or may be an account of a real journey – the documentary will work around the idea of exploration. Travel documentaries often take this form.

Fly-on-the-wall documentaries study a particular situation in very great detail – for example, family life. They use a mixture of techniques, including interviews and montage, to create meaning. Reality TV is a development of this form of documentary.

Docusoaps are a **hybrid** form of documentary, making use of codes and conventions of fly-on-the-wall documentaries and of soap operas to study a particular society rather than a small social group. They use soap opera conventions, such as the way characters are presented as **stereotypes**, to create meaning for the audience.

Docudramas are also a **hybrid** form of documentary, presenting a dramatic retelling of a real event. Some historical documentaries are presented as docudramas, to make them more immediate for the audience.

Documentary films, then, are **constructed** – they do not just happen. The documentary maker will have decided what versions of reality we should receive. As an audience, we must deconstruct this construction to receive the meaning.

Early documentary makers such as **Robert Flaherty** (*Nanook of the North*, 1922) and **John Grierson** (*Night Mail*, 1936) established a strong documentary tradition in this country. They saw the documentary film as 'the creative use of actuality'. It had to be useful, to do good. 'I look on cinema as a pulpit,' said Grierson, 'and use it as a propagandist.'

Flaherty was the first documentary maker in this country to employ the techniques of fiction when creating a documentary. He introduced the concept of 'emotional engagement' with the focus of the documentary and allowed the film to tell a story.

Checkpoint 2

What sort of documentary is *Big Brother*?

Action point

Would you say that lifestyle programmes such as property programmes are true documentaries or that they are aspirational entertainment programmes?

Example

The first docusoap in the UK is generally considered to be *Airport*, first broadcast in 1997, which centred on the lives of people working at Heathrow airport.

Exam question   answer: page 82

Look at the listings for a week's television programmes in a TV listings magazine. Identify how many different documentaries are being shown by each of the main terrestrial channels and what types of documentary they appear to be. What conclusions can you draw about popular types of documentary and audience expectations on the different channels? (1 hour)

Advertising and marketing

Checkpoint 1

What is a lifestyle magazine?

Check the net

Most of the big advertising agencies have their own website with an archive of ads they have produced. Large institutions such as Guiness or Levi's offer their own advertising archive as well.

Checkpoint 2

What is a USP?

Advertising is a fundamental part of our lives now. Wherever we go or whenever we consume any form of media we are subjected to advertising in some way. If we travel about, we are faced with billboard adverts by the road, adverts on buses or on posters at a station. When we open a newspaper or magazine there are adverts on many of the pages. If we watch commercial television, there are advert breaks every few minutes. Even if we limit ourselves to watching only the BBC, we are faced with adverts for forthcoming BBC programmes or for BBC publications.

Advertising

Adverts have four basic characteristics:

→ Adverts are commercial.
→ Adverts have an ideology.
→ Adverts aim to 'sell'.
→ Advertisers seek large audiences.

If you pay 10p to put an advert in the window of the local shop, to sell some of your old CDs, you are advertising them to an audience. It's exactly the same process when Skoda promotes a new car on the television. Skoda creates a certain image for its new car visually in the hope that you will go to the showroom and buy their car because you like the image they have constructed.

The key is that you have not bought a car but an **image** – i.e. an idealised image of yourself and your life. If you buy this car, drink this beer, wear these clothes, your life will be more like the ideal lives being shown in the adverts. So you are not buying the product but **aspiring** to the **lifestyle** being offered. You are buying the particular **benefit** the product can have for you, not the product as a 'thing'. Advertisers call this the **USP** (unique selling point) of the product – in other words, what is so unique about this product and what benefit it will bring you when you buy it.

Over 65 per cent of the money spent on advertising each year is spent on advertising in the press (national and regional press, magazines, journals and directories). TV accounts for about 25 per cent of the total figure. This is primarily because advertisers can target their audiences more directly by advertising in **niche** publications. As you can see from these figures, much less money is spent overall on advertising such as billboards because they don't reach specific audiences in the same way.

AD formats

In general, adverts follow four basic formats:

→ **Product-information format** – The product is the centre of the focus and its virtues are pointed out and explained, as is common in washing powder adverts.

→ **Product-image format** – The product is associated with images you may not readily associate with it, as in car adverts.

→ **Personalised format** – the product takes on human qualities, as with PG Tips or Tango. These campaigns can be long-running and very successful.

→ **Lifestyle format** – the product is associated with a particular lifestyle, as with mobile phones or alcohol adverts.

Advertising techniques

Within these ad formats, advertisers use several techniques to fool us into buying products:

→ **Flattering words** – where empty but colourful words are used such as: fights, improved, new . . .

→ **Celebrity endorsement** – a celebrity tells us how wonderful a product is – if it's good enough for them . . .

→ **Expert endorsement** or **statistical proof** – if an 'expert' says it's the best or if the 'statistics' prove that this is the best product (eight out of ten cats . . .)

→ **The new ingredient** – a new ingredient or a changed formulation means the product is better than ever before!

→ **Nostalgia** – reminds us of how much better things used to be – and the product harks back to these 'good old days'.

→ **Lifestyle fantasy** – if the consumer only had this product, their lives would immediately improve!

→ **Family values** – this ready meal or this drink will enable the whole family to enjoy idealised family meals or outings.

→ **Happy kids** – if you buy this toy for your children, they will be happy for ever. (N.B. Remember that adverts aimed at children 'teach' them that this is the case.)

→ **Selective advertising** – editing can make a product look good! It would be no good showing the ten shots of a cat turning away from the food bowl, for example.

→ **Music and sound effects** – the right soundtrack will establish the right image. This is why many adverts use well-known songs because they can sell their product by associating the product with the ideology of the advert.

Of course there are other techniques but it's important to see how common all of these are. How many examples of each type can you think of? Which are more successful on television and which in print? Why is that?

Check the net

Your school or college should be able to get access to www.creativeclub.co.uk which has access to all contemporary UK ads.

Examiner's secrets

When you are analysing an advert a good starting point is usually the ideology – what benefit from this product (e.g. happy family life) is being offered to the audience?

Exam question   answer: page 82

Buy a copy of a lifestyle magazine. How many adverts are there in the magazine? What USP is established for each product being advertised? What is the format and technique used for each advert? How do these adverts relate to the target audience for the magazine?   (1 hour)

Magazines

Magazines are among the most popular media texts in contemporary society and there is an ever-increasing number of niche magazines being produced. Magazines have very specific audiences (niche audiences) and magazine genres are clearly differentiated (men's lifestyle magazines are very different from magazines for pre-school children, for example) even though they all use the same codes and conventions.

Magazine codes and conventions

There are certain features which are common to all magazines.

Advertising is a very important source of finance for all magazines. Most magazines have:

→ an advice column
→ a contents page
→ a letters page
→ some form of makeover article
→ competitions
→ an 'In the next issue' page.

They also generally have merchandising or review sections. Review sections are usually a slightly more covert form of advertising since manufacturers will give the review items to the magazine free in return for their recommendation in the magazine.

Magazine front covers also always have strong generic elements:

→ price, bar code and issue number/date
→ a photograph directly related to the feature article
→ a recognisable masthead
→ various plugs and puffs on the front cover to entice the audience to buy the magazine
→ a thematic link between the colours used and the month of publication as well as colour links between images and text, for example.

Magazine front covers are carefully designed to attract the target audience and be easily identified on the newsagent's shelves. It is generally very easy to identify the genre of a magazine and the likely audience from a quick glance along the shelves.

Magazine institutions

There are a lot of magazine publishers in this country, some of whom are also part of larger media organisations. Just as with washing powder, there are many competing titles produced by the same institution. Although this may seem strange, if you think about how many magazines are published and how carefully each one is targeted at a **niche** market you can see that publishers do this to meet the needs of these niche audiences. By creating a range of closely related titles, of course, the publisher is seeking to persuade the audience to buy more than one and to attract very loyal niche audiences which ensure the

Action point

Try to look at all the magazines in a major newsagent's every month. Notice the similarities and differences in the front covers of magazines in the same genre and the way front covers vary at different times of the year.

Checkpoint 1

What do we mean by house style?

Don't forget

You can usually find out a lot of the institutional information you need by looking on the contents page or inside back cover of the magazine.

survival of the title. They can also establish brand loyalty by employing the same **house style** across their titles.

EMAP publishes a range of consumer magazines (such as *FHM*, *Empire*, *J17*, *Smash Hits* and *Internet* magazine as well as a range of B2B magazines (trade magazines).

IPC Media publishes titles such as *NME*, *Mizz*, *Essentials*, *Now*, *Nuts*, *Woman's Own* and *TV Times*.

Dennis Publishing publishes titles such as *Computer Shopper* and *PC Pro*.

Future Publishing publish a very wide range of titles such as *Cross Stitcher*, *Hi-Fi Choice*, *Internet Advisor*, *Net Gamer*, *Mountain Biking UK* and *Total Film*.

Haymarket Group publishes *Four Four Two*, *Revolution*, *F1 Racing* and *What Car?*

The **BBC** publishes a range of magazines as well, many related closely to the television programmes they broadcast, such as *BBC Gardeners' World* and *BBC History Magazine*. It also publishes other titles such as the *Radio Times*, *Noddy* magazine and *Eve*, for example.

Other types of magazine

There are also other types of magazine, such as the **in-house** magazines now produced by large retailers, such as Sainsburys and Tesco, as well as those produced by travel companies such as Virgin or British Airways. These titles are slightly different from the more conventional titles since they are often not sold but distributed free in appropriate locations, e.g. in an aeroplane. Once the magazine has reached a certain status and audience position, the publisher can then start to charge the audience – which adds to the credibility of the magazine and helps support high production values in itself!

Magazines and advertisers

One of the main reasons for the increased **segmentation** of the magazine market has been the drive from advertisers for magazines to closely target particular audiences. By establishing a **niche** market, the magazine is well placed to advertise products directly to the right audience, rather than the advertiser running a broad spectrum campaign. Far better to spend the money advertising a particular type of maggot for catching trout in 'Trout Catcher magazine' than spend money on a larger campaign which may not reach the right audience!

This is another reason why magazines are often **aspirational** – especially **lifestyle** magazines. It is encouraged by advertisers who are prepared to pay more to advertise their products in a **niche** magazine and are keen to see that magazine promote their products as part of an ideal lifestyle to aspire to.

Checkpoint 2

What is a niche audience?

Example

Tesco produces a range of in-house magazines targeted at different sections of their audience, e.g. mothers with babies and vegetarians. All of these promote Tesco products and the ideologies are very aspirational.

Exam question   answer: pages 82–83

Study the front covers of three magazines published by the same company. What elements of house style can you see across all three titles? In what ways does each one target a niche audience?   (1 hour)

Radio

Checkpoint 1

What is narrowcast broadcasting?

Watch out!

Remember that when you are writing about radio you must remember to write about the context and content of the radio programme, not just deconstruct the use of the four codes.

The BBC first started broadcasting in 1922; Radio Luxembourg started in 1933 and in 1939 the Home Service was launched. Pirate radio started in 1964 with Radio Caroline, broadcasting from a ship in the English Channel. It was so successful that the BBC launched Radio 1, 2, 3 and 4 in 1967. BBC Local Radio was launched in 1970 and commercial radio stations were able to start broadcasting locally in 1973 and nationally in 1992 with Classic FM. The Broadcasting Act (1990) paved the way for a much more flexible UK radio landscape which also made it easier for small, local radio stations to get licences. RAJAR was established in 1992 to record listening figures across national and local radio. 2002 saw the establishment of the latest BBC Radio stations, such as BBC Radio 6 Music, and progressively more radio stations are streaming their programmes globally across the Internet.

The codes of radio

The medium of radio is known as an 'intimate' medium because it seems to be broadcast directly to you, rather than a mass audience. We often listen to the radio in the background while we are doing other things. Because of these two factors, it is very important that a radio station has a very clear identity that is established within seconds of listening to it. We have to be able to identify the genre and content of the station very quickly to establish whether we wish to listen to it. Local radio in particular can be very narrowcast in terms of audience, broadcasting to a very specific local audience.

Radio has four main codes, which are usually defined as being:

→ words
→ sounds
→ music
→ silence.

Since radio can only communicate by using sound or lack of sounds, we can deconstruct a radio programme by identifying the combination of these four codes used.

Deconstructing radio

When studying radio broadcasting, you need to think about the institutional context of broadcast, the use of presenters or DJs and the way the programme is being presented. All your analysis is likely to centre on how the ideologies of the institution and the programme are established by these four codes. Remember that programmes at different times of day on the same station may have different ideologies because radio audiences vary at different times of day – even more so than television audiences.

You need to consider:

→ Who has made the programme? You would expect differing ideologies and target audiences from the BBC than from a small, independent local station.

→ Who is presenting the programme? A radio presenter or DJ has to establish a very clear identity if the audience are to become fans of their show. Different radio presenters do this in different ways but most of them use jokes and interactivity in the form of phone-ins and competitions, as well as trying to communicate their personality as they present.

→ How is the programme constructed? Is it a commercial station where the programme is regularly interrupted for advertising? If so, is this preceded by a station jingle, for example? Are there different elements in the programme, such as frequent news headlines, weather reports or traffic reports? How do these affect the overall shape of the programme?

Radio institutions ●●●

Obviously, the largest radio institution in this country is the BBC, which currently broadcasts many radio stations both nationally and internationally. These include Radio 1–4; the BBC World Service; Radio 5 (talk and sports); 6 Music; 1xtra and BBC Asian network; two ethnic stations and BBC7, which is based around comedy, drama and children's programming. National commercial stations include Classic FM, Jazz FM, Virgin and PrimeTime Radio among others.

Radio and new technology ●●●

Digital broadcasting is changing radio broadcasting, just as it is changing television broadcasting. DAB (Digital Radio) is increasing the number of stations broadcasting as well as improving the quality of reception. At present, all radio stations are free to air but it may be that the technology will develop in the future so that you will need to subscribe to digital radio stations just as you currently need to subscribe to most cable and satellite television channels. This may well affect the quality and variety of radio stations available.

Checkpoint 2

What is a jingle?

Checkpoint 3

What is a station ident?

Check the net

Digizone is a new interactive radio service, with additional resources available to listeners of stations such as Classic FM on their PC. Convergence is becoming an important aspect of all media forms. Look at http://www.thedigizone.co.uk/digizonesite.html

Exam question answer: page 83

Compare and contrast the programmes available at a particular time of day (e.g. 8 a.m. or 5 p.m.) across three different radio stations. How does each establish a particular identity and audience? (1 hour)

Newspapers

Despite the falling sales of newspapers in this country, they remain an important media form for study. For many years, newspapers were the only means of news reception available to audiences and, even now, the newspaper market is still significant. Given the rise in online editions of most of the popular newspapers, you should be aware of the basic forms of newspaper and the significant institutions who produce them.

History of newspapers

There were forms of newspapers in many cultures, and among the earliest is believed to be Berrow's *Worcester Journal*, which started life as the *Worcester Postman* in 1690. It was published regularly from 1709 and is believed to be the oldest surviving English newspaper. The first daily newspaper was the *Daily Courant* which started in 1702. In 1785 John Walter founded the *Daily Universal Register*, which became *The Times* in 1788 and is the longest-running newspaper in this country.

Improvements in printing technology during the nineteenth century, the establishment of news agencies to collect global news (such as Reuters) and improving literacy levels led to the establishment of more newspapers. In 1903, Alfred Harmsworth founded the first tabloid, the *Daily Mirror*. The Press Association was established in 1868, to monitor standards in newspaper publications.

The *Guardian* started life as the *Manchester Guardian* and became a daily newspaper in 1855 as a left-wing alternative to the *Daily Telegraph*. It became the *Guardian* in 1959. The *Sun* was launched in 1964 and acquired by Rupert Murdoch's News International in 1969, along with the *News of the World*. He also acquired *The Times* and *The Sunday Times* in 1981.

In recent years readerships have declined as audiences have received their news from news broadcasts and online news services rather than newspapers, but newspapers are still a very important area of study. Each takes care to establish a very particular ideology within contemporary society to attract a particular audience. Newspapers are frequently our benchmarks for moral attitudes and responses to particular situations because of these explicit ideologies, which are stated far more strongly than in broadcast news.

Broadsheet v tabloid

Traditionally in the UK there have been two types of newspaper:

→ **broadsheet** papers such as *The Times* and the *Daily Telegraph* and *The Financial Times* which focus on 'hard' news such as international news stories, political and economic issues, reporting in a factual way with a high level of discourse

→ **tabloid** newspapers which are smaller in size (easier to hold), with more emphasis on 'soft' news stories (e.g. celebrity news stories) and a far greater number of photos on each page.

Recently, this has changed again as some of the **broadsheet** newspapers are changing to become compact editions, i.e. tabloid size but still retaining the ideologies and expectations of a broadsheet newspaper.

It is important to start any analysis of a newspaper by identifying whether it is a broadsheet or tabloid publication, as this will directly indicate the ideologies, news coverage and context of production for the newspaper. Broadsheets traditionally have a very stable readership who read the paper every day, so can concentrate more on **stories in depth**. Tabloids tend to depend more on attracting **floating** readers – hence the emphasis on strong images and dramatic headlines.

Local and national newspapers ●●●

Local newspapers have to fight hard for **market share** against the nationals. Most are now weekly and many belong to one of the main newspaper groups. This means that you can see a **house style** across many different local newspapers, and journalists may contribute stories across several regions. Most local newspapers try to keep a local **identity** because they know that is what their audience want. They will often mimic the layout of national tabloid newspapers to attract audiences. Many local newspapers are becoming tabloids, with a much greater emphasis on community stories and less on national stories, in order to ensure they can secure their market share with a particular, local identity in this very congested marketplace.

News institutions ●●●

There are two main newspaper institutions in this country.

→ Associated Newspapers (chairman Lord Rothermere) owns the *Daily Mail*, the *Mail on Sunday*, the *Evening Standard, This is London* etc, as well as very many local newspapers, Teletext, GWR Radio, Independent Television News and British Pathé, among other institutions.
→ News Corporation (which owns News International), owned by Rupert Murdoch, publishes the *Sun*, the *News of the World, The Times, The Sunday Times* and many other newspapers worldwide. It also owns Sky TV, BskyB and Fox TV, film and cable institutions.

Other national newspaper ownership breaks down as follows:

→ The *Daily Express, Daily Star, Sunday Express* and *OK! Magazine* are owned by Northern and Shell
→ The *Daily Telegraph* and the *Sunday Telegraph* belong to The Telegraph Group
→ The *Independent* is owned by Independent News and Media PLC
→ The *Guardian* and the *Observer* is owned by the Guardian Media Group which also owns other institutions such as Jazz FM and *Auto Trader* magazine.

Action point

Reading the editorial section of a particular newspaper regularly will help you identify the position of that newspaper as the editor is often quite outspoken in their views.

Checkpoint 2

Tabloid and local newspapers aim to create an emotional response in the reader. Why?

Don't forget

News International is part of a much wider global organisation called News Corporation.

Exam question   answer: page 83

Look at the reporting of one particular news story across at least three newspapers. In what ways does each institution clearly show its differing ideologies?   (1 hour)

News broadcasting

Along with newspapers, news broadcasting is a very important media form to study. The news broadcasters exert a significant influence over the reception of news and social attitudes, although in a less explicit way than the newspapers. Digital channels have given us access to global news and many news agencies, such as Reuters, have realised the important role of news broadcasters in today's society and have started to broadcast direct to their audience rather than through news institutions. News is becoming a key commodity to be transmitted and controlled by powerful institutions.

Checkpoint 1

What is a news agency?

BBC news

Although when the BBC first started broadcasting it could not broadcast news because of fears from newspapers that they would be put out of business, BBC news soon became a key element of the Corporation's output.

Today, the BBC broadcasts news in a variety of ways:

→ frequent news bulletins on BBC One
→ rolling 24-hour news on News 24
→ news bulletins for all BBC radio stations
→ regional news programmes for each BBC region
→ constantly updated news on BBCi
→ BBC World Service News, broadcasting globally.

The BBC is not only a news broadcaster but also a news agency, employing reporters and finding news as well as reporting it. It has control of the whole process of news production and broadcast, unlike ITN, which is a news agency, providing news for a range of news broadcasters who do not also act as agencies.

Watch out!

The TV News franchises are reviewed regularly. Sky News has gained some of the franchises in recent years so you should make sure you know which organisation supplies each news broadcast which you study.

ITN

ITN (Independent Television News) is now one of the largest news organisations in the world and was founded in 1955, as an independent organisation owned by the ITV companies, producing news programmes for broadcast on the ITV network. It operates as a news agency, providing news for these different broadcasters to transmit.

ITN's output includes:

→ lunchtime and evening news for ITV and Channel 4
→ ITV News Channel
→ Channel 4 News
→ *First Edition*, a weekly news and current affairs programme aimed at nine- to thirteen-year-olds
→ Five News with a fast-moving format specifically aimed at younger audiences
→ ITN Radio provides the news service for Independent Radio News (IRN), transmitted by 260 independent local radio stations
→ ITN Radio supplies news to Classic FM, Talk Radio and Atlantic 252, LBC 97.3FM and LBC News 1152AM.

Digital news services

There is now an increasing range of global news agencies and institutions which offer news in the UK, including the Reuters agency, CNN, Fox News and Sky News.

Reuters

Reuters was started in October 1851 and was soon collecting and transmitting news around the world. It is now the largest multimedia news agency globally.

During both World Wars, Reuters came under pressure from the British government to serve British interests so it became a private company, to preserve its independence and neutrality. Reuters stipulates that its integrity, independence and freedom from bias must be upheld at all times. It is successful because of this neutrality (for instance, during the Iraq war, only Reuters reporters were permitted to report from many Iraqi-held areas).

Recent technological developments have enabled Reuters to broadcast directly, e.g. with video streams from its website, whereas previously its material was only available to news institutions. These developments are blurring the boundaries between news agencies and news broadcasters still further.

CNN

Cable News Network was founded in 1980 by Ted Turner. It is currently owned by Time Warner. CNN started 24-hour news coverage. It now includes 15 cable and satellite television networks globally, 12 websites, and two radio networks. CNN is available in Britain to all digital subscribers.

CNN's saturation coverage during the Gulf War led to accusations that it was a propaganda outlet for the US. During the Iraq war, the Iraqis insisted that other news agencies could not share reports or images with CNN for this reason.

Fox News

Fox News Channel is owned by Fox News Network and is often considered the most important US cable news channel. Fox News Network is owned by News Corporation. It is perhaps best known for news personality Bill O'Reilly and because it has trademarked the description 'fair and balanced'.

Sky News

Sky News was Britain's first dedicated 24-hour TV news channel, launched in 1989. It is now respected as offering good-quality news coverage and maintaining a clear identity in an increasingly competitive marketplace. It is part of BskyB and is still partly owned by Rupert Murdoch's News Corporation. It now offers Sky News TV, Sky News Radio, Sky News Online and a range of other news services, in line with other Sky channels and other news organisations. As part of News Corporation it has access to a very large news agency to generate the news for broadcast.

Check the net

You can visit the Reuters website at www.reuters.com and view live video footage and get live news.

Checkpoint 2

What is saturation coverage?

Exam question   answer: page 83

Record at least three different TV news broadcasts for the same day and roughly the same time (e.g. early evening). What can you say about the way the news is packaged by each of the different broadcasters?   (1 hour)

Video/computer games

Video games probably began in 1958 when *Tennis for Two*, a precursor to *Pong*, was developed. Early games were almost all played on computers, consoles not being as sophisticated. The first commercially successful game was *Space War!* in the 1960s.

The first console video game with widespread success was Atari's *Pong*, developed in 1972. The game was similar to tennis. The first game to cause controversy was probably *Death Race*, produced by Exidy in 1976. Killing gremlins by running over them with a car was felt to be unnecessarily violent.

Pacman was released in 1980 and, simple as it was, became one of the most engaging games ever released. Other key games in the development of the industry were Nintendo's *Super Mario Brothers* (1985) and *Sonic the Hedgehog* (1991). At this point, the market changed because consoles were released and games began to be created for different platforms.

Mortal Kombat was released in 1992 and, as well as being important because it was the first game to create characters as we know them today, it was heavily criticised for its violence.

In today's market, there are games for all the different platforms – PC, Xbox, Gameboy and Playstation – and a wide variety of games available.

Checkpoint 1

Are all games available on all platforms?

Check the net

A good site to start with when you are researching computer games is http://www.game-culture.com/.

Game genres

As the gaming industry grows increasingly more sophisticated, it is becoming easier to identify the central game genres and therefore the codes and conventions associated with those genres. These are tending to develop into two types of game – 'gamer' (more violent) titles and 'consumer' (softer) titles. Among these are:

→ **Adventure** – Adventure games cast the player as hero. There is usually a puzzle to solve and a treasure to find. Adventure games are usually narrative-driven and often related to film releases and television programmes, enabling the player to take on a ready-made identity.

→ **First person and other 'shooter' games** – FPS games cast the player as protagonist (or occasionally antagonist), holding the weapons and fighting, with good reflexes. Textual codes will usually create a complex environment and ensure high levels of player participation. More generic shooters do not involve the player constructing an identity, simply 'shooting', e.g. *Pacman*-type games.

→ **Roleplay and strategy games** – Roleplay games employ sci-fi or fantasy codes and conventions. They have developed from roleplay games such as *Dungeons and Dragons*. Players take on a particular character (usually from a specified range) and this character will gain skills and/or tools as they progress through the levels of the games. Strategy games build from this, involving more than one player at a time and turn taking. Online wargames such as *America's Army* are good contemporary examples of these games.

- **Simulations** – Simulations require the player to learn a particular skill (e.g. fly an aircraft) or build something (*Sim City*) and usually come with detailed manuals. They have high levels of interactivity and feedback for the player and involve planning and strategy.
- **Sports and racing** – Some sports games play the sport, others are strategy-based. This genre is extremely competitive, just like real-world sports. Racing games are also traditional. The player is in the driving seat of a high-performance vehicle and competes against other drivers or sometimes just against time.

Changes in audience profiles ●●●

Whereas once the primary audience for games was seen as male, aged 11–15, the gaming markets are expanding, due to a number of factors:

- More sophisticated graphics and more complex game play have attracted more mature audiences.
- Early gamers continue as gamers – those who were in the original audiences in the 1980s are older but continue to play.
- Industry growth has enabled increasing diversity of games, leading to broadening of audiences.
- Industry growth has meant that games are now cheaper to buy (although more expensive to produce), which again broadens the audience.
- Gaming is becoming culturally acceptable across many social groups.
- Online gaming has considerably expanded game appeal – for example, increasing female audiences, who are participating in traditionally 'male' games as well as more 'female' games such as simulations.

Checkpoint 2

What is a 'clan'?

Comments on and criticisms of computer games ●●●

- Game play can be competitive, co-operative or individualistic.
- Many games are violent. Others are to do with excelling in sport, completing dangerous missions to retrieve or collect things, or taking on the persona of a warrior or hero and employing strategy to win. Games use technology to represent reality or to embody fantasy. Critics argue they harden players to the task of murder by simulating the killing of hundreds or thousands of opponents in a single game. *Grand Theft Auto 3* is criticised, for example, because players can choose to steal a car, pick up a prostitute, have (implied) sex with the prostitute, then kill her and steal her money.
- *Tomb Raider* was marketed through the visual construction of Lara Croft.
- Many critics have commented on the lack of 'female' games.
- Critics also observe how few original games are produced, since so many games are spin-offs from films.

Action point

When researching a game, look closely at its packaging and watch the opening sequence very carefully. How do they set up the genre and expectations for the game?

Exam question   answer: page 83

Choose one game with which you are familiar. What codes and conventions are used in the game? Is it typical of its genre? In what ways does it construct its target audience?   (1 hour)

Answers
Media forms

Commercial television

Checkpoints

1 A star vehicle is a programme or film which is written for a particular star. It may be, for example, that Granada will have 'bought' a particular star for three series and will then set about designing three series for them to star in.

2 Dumbing down is an accusation often made against broadcasters and publishers that they are reducing the quality of programmes to the LCD (Lowest Common Denominator) of the media. It is claimed that they are interested in producing simple television for mass audiences very cheaply and not concerned with offering quality programming or publishing.

Exam question

A strong answer here will take a broad look across the different channels for one day and draw a series of conclusions. It does not really matter whether the conclusion is that television is dumbing down or that quality standards have not dropped, as long as the answer is well structured, with good use of examples.

A secure answer would be able to position commercial television clearly against the competition of PSB and may well make reference to the accusations of dumbing down which have also been made against the BBC. It would be logical to break the commercial channels into groups for this response in order to avoid having to identify each channel in turn. So, for example, one group might be the free to air commercial channels, another group might be the narrowcast channels, another group might be the film channels or the Sky channels. By then looking at particular key pivot times (where the maximum audiences are targeted) such as 8 a.m., 7 p.m. and 9 p.m., conclusions can be drawn about what programmes are going 'head to head' with those on other channels and what quality choices are available for viewers.

Confident candidates might well be able to extend this debate to consider the differing situation for daytime television and the involved debates surrounding audience uses and gratifications.

Public service broadcasting (PSB)

Checkpoints

1 The licence fee is a legal requirement for any household in the UK which owns even a single television. There is no additional charge for more than one television. There are reductions in the licence fee for OAPs. It does not matter if you do not wish to watch BBC channels, you must have a TV licence to view free to view television.

2 As part of the PSB requirements for the BBC, it does not currently offer any paid-for channels. The terrestrial channels and the digital channels on offer are all free to view.

Exam question

The issue of the licence fee continues to be debated in many places. Commercial television channels obviously resent the BBC being funded and therefore not required to generate revenue from advertising sales. The BBC argues that the licence fee is essential to support its PSB remit and that if the licence fee were taken away it would not be able to provide the range and calibre of programmes that it currently provides. Yet Channel 4, for example, also has a PSB remit but no licence fee funding.

There are wider debates to include here, however. For example, BBC Worldwide is the publishing arm of the BBC and generates extensive amounts of revenue for the BBC by selling BBC programmes abroad and marketing BBC magazines, etc. This money is in addition to the licence fee and probably far exceeds it.

Another key debate centres around the changing face of broadcasting as digital television takes over. The analogue television signal (the terrestrial channels) is due to be turned off in approximately 2010 and from that point all television signals will be received through cable or satellite. In preparation for this the BBC has extended its provision of free to view channels to include the new digital channels such as CBBC and BBC Three. However, households in some parts of the country are not able to receive these channels – yet still have to pay the full licence fee.

Post-Hutton, of course, stronger candidates may also debate the freedom of the BBC and its perceived integrity gained from the status given by the PSB remit and consider whether this is now inappropriate.

Reality TV

Checkpoints

1 Cinema verité is a type of documentary which attempts to produce film which is unmediated – i.e. unedited – and which does not rely on carefully framed shots and deliberate camera angles to create meaning. There is more about cinema verité on pp. 122–127.

2 A C list celebrity is someone who is not a very well-known celebrity – for example, a game show contestant who has become famous simply by being a game show contestant. A list celebrities are 'true' celebrities.

Exam question

Obviously, the structure of this response will depend on the particular texts chosen for exploration but, unlike some of the other questions in this section, such an answer should revolve around very detailed textual analysis.

Candidates should be able to contextualise their study in relation to the development of the genre, although this will probably only be a brief referencing. A useful starting point here might be the debates raised by the different series of *Big Brother* and the way that ordinary people were cast into certain roles and stereotyped in specific ways. The editors usually have 24 hours' worth of material to choose from when editing for any one highlights show and so it is pretty easy for them to construct a particular representation for any one character.

Issues arise around related areas. For example, in most of these programmes, the contestants are isolated from real life, so have no concept of how they are being represented and therefore no chance to defend themselves. In celebrity game shows, it might be argued that the celebrities should be used to how the media construct representations and

therefore able to behave in such a way that the representations constructed will match with their intended personas. However, ordinary people do not always have those skills and many contestants have complained subsequently about the way that they were portrayed.

There are also interesting debates here relating to audience expectations. Given that audiences conventionally expect conflict, there must be 'nice' characters and 'nasty' characters. Producers of such shows argue that audiences seek to create opposites in this way whether they are offered to them or not, so they are simply responding to the audience's expectations. Indeed, in some cases they argue that they use feedback from the audience before deciding how the contestants will be cast.

You might also want to consider the words 'character roles' in this question and the ideologies they imply for the programmes. Do the programme makers keep the programmes 'real' or do they create a 'story', for example?

Conclusions to this essay therefore will be based around the degree to which the contestants are manipulated by the programme makers and the relationship between the programme makers and the audience. Do the programme makers make the programmes that the audience demand or are the audience also constructed by the ideologies presented to them by the programme makers?

Situation comedy

Checkpoints

1 Most sitcoms are recorded in front of a studio audience because the actors find it easier to work with an audience. Sometimes, however, they are recorded in a studio and a separate canned laughter track added during editing.
2 *The Simpsons* is sometimes described as a postmodern sitcom because it uses the sitcom formula for the cartoon in terms of basic structure but subverts it in many ways, such as the stream of celebrity guest stars who appear in different episodes.

Exam question

This is a relatively straightforward question since it is possible simply to identify a contemporary British sitcom and examine it in relation to each of the generic codes and conventions identified. Each point should be supported by detailed evidence from the text and you should make sure to address all the key concepts, not just focus on media language.

You could expand this essay a little at the end if you had time by exploring the sitcom in relation to its wider contexts of production and consumption. For example, you might consider the ideologies added by the scheduling of the programme and the likely relationship with the cultural context of the target audience or the validity of the representations in relation to this context.

Soap operas

Checkpoints

1 Metanarrative is the overall narrative for the soap opera, comprising all the interlinked subnarratives.

2 The locations for UK and US soaps are often very different because they are usually centred on close communities. In the UK these are traditionally found in working-class neighbourhoods, whereas in the US, community is seen as relating to family values shared by people who are often spread across a wide geographical area.

Exam question

On one level, like the sitcom question, this is fairly straightforward, simply requiring you to identify a particular UK soap and analyse it in terms of how it conforms to or challenges typical soap opera conventions. Each point should be supported with detailed evidence and the discussion related to the key concepts. However, the more challenging part of the question is the second part. The metanarrative of a soap opera is, by definition, never-ending, so each individual narrative must reach some kind of conclusion, yet not conclude the metanarrative, and each overlapping story must progress the metanarrative as well as its own storyline. The basic question here, therefore, is about whether this is a deliberate device – since there is no resolution, it is difficult for an audience to disengage from the programme. Coupled with careful scheduling to attract the (primarily) female audience, the metanarrative keeps the audience watching.

This can lead you to debates about audience expectations for the soap opera – is the fulfilment in identification with particular characters or is it, in fact, in engaging with these multiple storylines and remaining hooked on the developing events? Is this why soap operas, while supposedly highly realistic, can use devices such as long-lost children suddenly returning home from boarding school or a different actor/actress taking over a part without the audience objecting in any way?

Independent cinema

Checkpoints

1 Non-realist film does not pretend to be 'reality' and uses devices which highlight this fact, for example, flashbacks or characters speaking directly to camera.
2 Luhrmann used a lot of Bollywood codes and conventions in his musical *Moulin Rouge* (2001).

Exam question

There are many Bollywood movies available and obviously the choice of codes and conventions for deconstruction will depend on the chosen film – although, since the musicals tend to be quite formulaic, it should be fairly straightforward to identify them. They can be deconstructed in relation to generic conventions and also audience expectations. What ideologies are habitually constructed within these generic conventions, for example? The second part of this question asks you to think about how these codes and conventions have been used in Western films. Again, your choices will depend on the text you have chosen, but many contemporary films make use of Bollywood conventions. Increasingly, audiences outside India are becoming familiar with the genre and recognising these codes and conventions – an intertextuality which can add to the ideologies of the

film using these generic codes. As audiences become more sophisticated, they are starting to appreciate these intertextual links.

Mainstream cinema

Checkpoints

1 A CU is a close-up shot – usually just showing a head or a prop being used.
2 Vertical integration means that a single media organisation is responsible for all stages of production and distribution.
3 A multiplex is a multiscreen cinema, usually with very high-quality sound facilities and with often other franchises as part of the complex, e.g. a pizza restaurant. Most multiplexes are owned by the large film studios.

Exam question

Auteur questions are always popular at this level because they allow you to engage with a range of key concepts and demonstrate your understanding across a range of texts. To answer this question you need to have studied at least three films from your chosen auteur and to have done some research about their work. You should be aware of the chronological order of the films and the influences which shaped and formed each of them. Tarantino, for example, has talked about the way his film making has matured as he has become more subtle in his construction. This is the sort of analysis we would be looking to see supported by textual evidence and some thought about why he has made these changes as well as how and when. It is important to remember that an auteur does not work in isolation, so you need to look at the wider contexts of construction as well and consider how these have affected the films. There are also other issues to consider – when a film is made by an auteur, it carries a certain status and expectation. To what extent is that usually justified and to what extent does it affect the way the film is received? What advantages can there be for an auteur when constructing?

Documentary

Checkpoints

1 The London Marathon is usually filmed as a documentary because the focus is on the vast number of ordinary runners who take part, whether for themselves or for charity. It is not presented as a significant competitive sporting event.
2 *Big Brother* is probably best described as a reality game show. It is a game show because the participants compete to be the last person remaining but it is reality because we are shown real events. While the editors may select the material to show us, they cannot generate material or present participants inappropriately.

Exam question

The first part of this is fairly straightforward since you can basically complete a table to identify the types of documentary and their scheduling and base your answer on this table.

If this research is completed, the second part of this question is also quite straightforward since it will be reasonably clear what the popular documentary types are and when they are scheduled. You can then evaluate these findings in terms of audience expectations but also in terms of the role of television as an information medium. To what extent do you think the channels are seeking to fulfil audience expectations or are there other, institutional motivations for their programming choices? Are these documentaries likely to be cheap or costly television?

Advertising and marketing

Checkpoints

1 A lifestyle magazine is one where the ideologies are constructed around a concept of an aspirational lifestyle for the reader to aim for.
2 The USP is the unique selling point – the key feature about this text which makes it different from others and therefore more appealing.

Exam question

As with some of the other answers here, the logical starting point for this analysis is a table listing all the adverts in the magazine, the products being advertised, the USP for each advert and the format for the advert (look back to the section). Add a brief annotation about how these combine to reach the target audience (which will logically be the same audience as the audience for the magazine).

This is a fairly straightforward question if you have done your analysis first so that you can use it to support conclusions. You need to remember to keep including textual examples to support your points. Do not deconstruct each advert in turn in your answer because you will not have time. If you have already collated your conclusions by making notes in a table you will be in a much stronger position to write your answer in terms of the key concepts and not end up describing each advert in a list.

Magazines

Checkpoints

1 House style is the required format for the institution which produces the text – font size, typeface, layout, etc, are all defined so that all titles in the group to share a visual group identity.
2 A niche audience is a small audience – hobbyist magazines often cater for niche audiences, for example.

Exam question

Make detailed notes about each front cover, identifying the comparative features which are likely to be the house style and then identify features which attract a particular niche audience to the magazine. Once you have made these notes you will be able to identify which generic elements appeal to the audience and which particular elements appeal to the particular niche audience. To what extent are the house style similarities part of the generic codes and conventions of magazines and to what extent are they institutionally

defined? Does it matter? How do the magazines ensure that they can publish competing titles within the same genre? Why is it in the interests of an institution to do this?

Radio

Checkpoints

1 Narrowcast broadcasting is aimed at very small and particular audiences, e.g. a local audience which is small geographically and demographically.
2 A jingle is a very short snatch of song or music which is used in a particular radio programme to signify the programme. These are often created by the presenters.
3 A station ident is a short snatch of song or music which is used to signify the identity of the station as a whole and will be heard across all the programmes on the station.

Exam question

The best way to approach this question is by careful analysis before you begin your answer. Analyse each programme in terms of station, station identity, target audience, scheduling issues, institutional context and ideologies established. You will then be able to use your analysis to support your arguments about the identity which each programme establishes and the relationship it has with its audience. How important do you think it is that each programme manages to create this identity and ideology? Does this help retain audiences in congested airspace?

Newspapers

Checkpoints

1 The political affiliation of a newspaper is important because this will control all the representations and opinions in the newspaper. It is usually an important part of the ideology of the paper.
2 Tabloid and local newspapers usually deal more with 'soft' news and want an emotional response from the audience to keep them engaged with the paper. Broadsheets tend to offer more 'hard' news stories and are therefore less emotive in approach.

Exam question

Again, you need to complete your analysis before answering the question. By analysing the reporting of the story by three different newspapers (note: this may not be the same as three different institutions – an important point) you will be able to draw conclusions about the ideologies established by each institution and how these are conveyed to the audience. Why does each institution choose to report the story in this way? Is it a hard or soft news story? Does this affect the way it is reported? What representations are used by the newspaper? Why? What discourse is used? Why? Where is the story reported in the paper? Why? All of these factors are important in identifying the key ideologies and therefore the institutional values conveyed to the target audience.

News broadcasting

Checkpoints

1 A news agency is involved in collecting news but not usually distributing it. Some news agencies also act as broadcasters and in some countries the only news agency is controlled by the government.
2 Saturation coverage means that all coverage is devoted to one story. All the TV channels had saturation coverage of 9/11, for example, because it was so significant.

Exam question

This is similar to the previous question but phrased slightly differently because it is asking about news broadcasting, not newspapers. The basic analysis is still the same in terms of media language, discourse, representations, ideologies and institutional context, but the broadcast context affects construction and consumption differently from newspapers and the target audience is different. For example, newspaper audiences almost always agree with the political perspective of the newspaper they read, while news broadcasters have to reach an audience across the political spectrum. How do they manage to do this?

Video/computer games

Checkpoints

1 Most games are available for all main platforms – PC, Xbox, Playstation and often for Gameboy as well. The game play is sometimes different between different platforms and the release dates may vary as producers try to increase audiences.
2 A clan is a group or society of online gamers who have joined together to play a particular game as a team. Membership is organised in different ways by each clan for each game.

Exam question

This is again a fairly straightforward analysis question. Obviously, the codes and conventions which you identify and examine will depend on the game you choose, but the relationship between game and audience is usually pretty explicit for all computer games and the audience is assumed to be a competent audience who are familiar with these codes and conventions. Once you have researched your game and identified the codes and conventions being used you should be able to relate these to the target audience. The ideologies constructed within the game should appeal to the target audience and this analysis will enable you to evaluate the degree to which the game is typical of the genre, to what extent it constructs its target audience and to what extent it depends on an audience already constructed by other games. For this answer you should try to think about the wider contexts in terms of related texts as well – for example, many games are closely related to films and therefore inherit audiences and ideologies. So, is there a difference between these games and those games which have to construct their own ideologies, for example?

Revision checklist
Media forms

1 Understand PSB and the role of the BBC	Confident	Not confident. **Revise** pages 54–55
2 Understand how commercial television is structured and funded	Confident	Not confident. **Revise** pages 52–53
3 Understand the codes and conventions of sitcom	Confident	Not confident. **Revise** pages 58–59
4 Understand the codes and conventions of soap operas	Confident	Not confident. **Revise** pages 60–61
5 Understand the codes and conventions of documentary	Confident	Not confident. **Revise** pages 66–67
6 Understand the codes and conventions of niche magazines	Confident	Not confident. **Revise** pages 70–71
7 Understand the differences between tabloid and broadsheet and local and national newspapers and the organisations which publish them	Confident	Not confident. **Revise** pages 74–75
8 Understand the relationship of news agencies and news broadcasters and be familiar with the broadcasters in this country	Confident	Not confident. **Revise** pages 76–77
9 Understand the codes and conventions of radio and be familiar with the forms of radio	Confident	Not confident. **Revise** pages 72–73
10 Be familiar with Hollywood and the studio system and how this affects the context of production and consumption for blockbuster films	Confident	Not confident. **Revise** pages 64–65
11 Understand the context of production and consumption for independent film	Confident	Not confident. **Revise** pages 62–63
12 Be familiar with the main genres of computer game and the audiences for these games	Confident	Not confident. **Revise** pages 78–79
13 Be familiar with the rise of reality TV and the codes and conventions of different forms of reality TV	Confident	Not confident. **Revise** pages 56–57
14 Be familiar with the ways adverts are constructed	Confident	Not confident. **Revise** pages 68–69

At A2 level you are expected to be able to engage with a wide range of media issues and debates and these should be applied across all your written work. In this section we give you a brief introduction to many of the key theories and issues which you should be aware of.

You should remember that even for the synoptic units you are not expected to know a vast range of media theory or issues in great depth. It is the application of these theories to support your answers and/or understanding of the media issues and practice which will be rewarded, not simply whether you know the theories or media practice. So, you should practise as often as possible referencing a range of media theories in your essays, in relation to the texts you are analysing.

You should also check with your teacher which aspects of media theory they think you should know well, and also how you might incorporate some of these aspects into your written work in class.

Exam themes

→ Audience theories
→ Cultural theories
→ Representation theories

Topic checklist

○ AS ● A2	OCR	AQA	WJEC
Effects theory	●	●	○ ●
Reception theory	●	●	○ ●
Uses and gratifications	○ ●	○ ●	○ ●
Narrative theory	○ ●	○ ●	●
Representation and regulation	○ ●	○ ●	○ ●
News values	●	○ ●	○ ●
Audience profiling	●	○ ●	●
Advertising	●	○ ●	○ ●
Postmodernism	●	●	●
Genre theory	○ ●	○ ●	○ ●
Star theory and auteurs	●	●	●
Gender studies	●	●	●
Marxism	●	●	●
Distribution and exhibition	●	●	○ ●
Pluralism	●	○ ●	○ ●

Effects theory

Checkpoint 1

What is the culture industry?

The 'Frankfurt School' is the umbrella term given to a group of social scientists originally based at the Institute for Social Research, Frankfurt, in the years between the two World Wars, who were exiled to New York during the Second World War. They were concerned with the power which modern mass media had to propagandise on behalf of fascism. The founders of this school of thought were left-wing and they clearly articulated criticisms of a capitalist system which controlled media output, creating a mass culture that eliminated opposition or alternatives.

The Frankfurt School

The Frankfurt School was concerned with the impact of the rise of the media **industries** on society. Their argument was that the rise of the 'culture industry' resulted in increased standardisation within society. Under capitalism, society controls almost everything and even culture is processed through the mass media as something which is bought and sold. So culture is produced by the industry of media in just the same way as a car is produced by the industry of car manufacturing or a pair of shoes by the industry of shoe making.

The mass audience (it is seen as a passive 'thing', not a group of active consumers) is thus manipulated and indoctrinated by society (the **hegemony** of the ruling classes) and progressively less able to criticise it. The mass media prevent culture from being effectively communicated in any authentic form until it has first been **commodified** and changed to fit the capitalist system. So we buy fashion because the concept of original clothes has been turned into something which we can buy, believing that we are making a fashion statement, whereas all we are doing is buying an artistic vision which someone else has commodified as something good to buy! In terms of the media we consume, therefore, we watch TV programmes, not because we want to, but because we have been conditioned to watch them.

Links

You can find out more about hegemony on p. 12.

The hypodermic needle model

The original model proposed to explain how this worked was the hypodermic needle model, which demonstrates the effects of the mass media on their audiences. This model owes much to the supposed power of the mass media – in particular film – to **inject** the **passive** audiences with ideologies. This was most commonly seen in Nazi propaganda films such as *Triumph of the Will*. Repressive regimes across the globe control the media organisations in their countries, usually in the belief that strict regulation of the media will help in controlling their populations. For example, during the Iraq war, the only reporting of the war on Iraqi television was from the Iraq State News Agency – and not surprisingly their version of events was sometimes very different from that being broadcast outside Iraq. Saddam Hussein believed that if he told the people what he wanted them to believe then they would be **injected** with these ideologies and agree with him. The

Action point

There are film libraries where you can get access to some of these early propaganda films. Try the British Film Institute, for example, as it has a very extensive archive.

audience is seen as a **passive** mass who will immediately accept whatever version of events is given in the media.

Violence in the media ●●●

In our society, effects theory is most often cited as evidence of the dangers of violence in the media. Some of the 'moral majority', as they have been called, argue that TV output which is explicitly sexual, too violent or in other ways offensive must be **censored** as it may influence the audience to act in the same way. This assumes again that the audience is **passive** and will receive without question whatever ideologies are presented.

Although most people generally accept that today's audiences are more **media-literate** than this model suggests and are able to decode media texts without being injected in this way, this is still a popular argument for politicians and social commentators when **moral panics** are generated. Issues such as internet pornography or computer games allegedly damaging literacy skills or contributing to violent behaviour (e.g. *Grand Theft Auto*) usually centre around effects theory. Theorists will try to blame the media for all anti-social behaviour – which is usually a little simplistic. For example, the serial killer Jeffrey Dahmer watched a clip from his favourite film before every one of his murders. As you can imagine, effects theorists use this to justify the hypodermic syringe theory, but the film was *Star Wars* – clearly the film had very different ideologies for him than for most of us as receivers of the text.

Cultivation theory ●●●

Because it is difficult to prove the effects of individual media texts on an audience, a more refined version of the theory has been created called the '**cultivation model**'. According to this, while a single text does not have much effect, repeated exposure will make the audience less sensitive. Critics call it becoming '**desensitised**'. Theorists will cite the number of previously banned films which, when they are broadcast some years later, generate comments of 'I can't see what all the fuss is about', because social attitudes and expectations have changed so much.

Two-step flow ●●●

A final development from this theory which you should know about is the concept of **two-step flow** which refines this rather basic model a bit. Two-step flow assumes a more active audience, who will discuss media texts with each other. If the text is discussed with someone we respect (an **opinion leader**) then we may well be passive enough to accept their received views of the text. If this seems a little simplistic, just think of the number of TV critics in all the newspapers and how influential they can be about what we decide to watch!

Action point

Do you personally think there is too much violence on television?

Links

You can find out more about moral panics on pp. 140–141.

Checkpoint 2

What do we mean by 'desensitised'?

Exam question   answer: page 116

Looking at the televison schedules for this week, what evidence can you see to suggest that the mass media in this country still seek to promote hegemony?

(1 hour)

Reception theory

While uses and gratifications research is concerned with why people use media, the approach sometimes referred to as 'reception theory' (or reception analysis) focuses on what people see in the media, on the meanings they produce when they interpret media texts.

Watch out!

Remember that there is no single meaning for any media text. Meaning is constructed by the audience.

Action point

Which factors shape your responses to a media text most? Past experience? Liking a particular genre? Opinions of friends and family? Critics?

Checkpoint 1

What is the context of consumption?

Reception theory   ●●●

In a sense, this is an **extension** of uses and gratifications theory. Reception analysis concentrates on the audience itself and how it responds to the text. Reception analysis is based on the idea that **no text has one single meaning**. Instead, reception analysis suggests that the individual members of the audience **themselves** help to create the meaning of the text. We **decode** the texts that we encounter in **individual** ways – which may be a result of our upbringing, the mood that we are in, the place where we are at the time or, in fact, any combination of these and all kinds of other factors. So you may watch a television programme and enjoy every minute of it, while your friend may hate it. Indeed, we may well actually create a **different meaning** for it.

Reception analysis tries to understand these differences. Factors such as gender, our social status and our social context can be enormously important when we construct the meaning of a text. Of course, this kind of thing is often closer to psychology – the study of personality – than media studies and can be very difficult to research. While quantitative researchers simply count the number of people watching a programme, reception analysts have to make use of **interviews** in order to get some kind of idea of the meanings that people attach to texts.

Drawbacks to reception theory   ●●●

Uses and gratifications theory has a simple list of four types of use for the media. Because reception theory concentrates on the individual it can never do this – we are all different and no one theory can comprehend that. This can be seen as a strength of the theory – that it takes into account the complexity of our response to the media. At the same time the theory has a weakness in that it often ignores the **context of everyday life**. Our responses to a particular text may also be framed by factors such as whether we were in a good mood when we consumed the text.

The **context of consumption** can cause further complexities for reception analysis. Going to see a film in the cinema is such a different experience from watching the same film at home that the meaning can be so different that reception analysts would suggest that it is almost a different film. However, the theory does not explain all of the contextual factors that make cinema so different from television.

Uses and gratifications theory looked at **why** we make use of the media, reception analysis looked at what we see when we watch a

media text – the remaining question is that of **how** the media fits in with our everyday lives – how do we live with the media?

David Morley ●●●

David Morley has explored the '**politics of the living room**' – the idea that the media is just part of all the different things that may be going on in your home. In a typical family it can also be a subject of argument or a symbol of power. This may be a concept that you will find quite familiar!

We can never consider one example of the media on its own – we are always choosing from many different alternatives and, more confusingly, our understanding of one text may be affected by our knowledge of another – we will have **expectations** based on our experience of similar texts, for example. It is also unusual to concentrate fully on any media text – we may skim read through a magazine or channel hop while watching a soap opera, for example. Most audience theory assumes that you have been concentrating – ethnographic theories accept that you may skim.

The media can become an important part of the routines of our lives – you may want to watch *Neighbours* when you get in from school or listen to the *Chart Show* every Sunday when you do your homework. In these examples, the way that the media text fits in with the pattern of your day is almost as important as what the media text is. It is also likely that you share your media consumption with other people – family or friends – who may influence your choices and your responses.

The nationwide study

In 1980, David Morley conducted a very detailed audience study, observing how many different social groups read the same media text. He discovered that there are three main types of reading for any media text:

→ **Dominant** (or **hegemonic**) reading – The reader shares the programme's code (its meaning system of values, attitudes, beliefs and assumptions) and accepts the **preferred reading**.
→ **Negotiated** reading – The reader partly shares the programme's code but modifies it in a way which reflects their position and interests.
→ **Oppositional** (**counter-hegemonic**) reading – The reader does not share the programme's code and rejects the preferred reading, bringing to bear an alternative frame of interpretation (e.g. a feminist reading of a 'lads' magazine).

Action point

Think about the ways in which you use the media – for example, listening to MP3s, watching soap operas or reading magazines. What influences your choices of what media to consume?

Checkpoint 2

What is a dominant, a negotiated and an oppositional reading?

Exam question   answer: page 116

Choose two programmes which you know well and identify dominant, negotiated and oppositional readings for each of these texts. Which different audience groups might make these readings?   (1 hour)

Uses and gratifications

What four motivations for choosing a text did Blumler and Katz suggest?

The opposite range of audience theories to effects theory is uses and gratifications theory. The basic difference is that this assumes an active audience, which chooses the texts it consumes and where individuals have differing reasons for consuming these texts.

Uses and gratifications theory ●●●

During the 1960s it became apparent that audiences made **choices** about how and when they consumed media texts. Far from being a passive mass, audiences were made up of individuals who actively consumed texts for different reasons and in different ways.

Blulmer and Katz expanded this theory in 1974, suggesting a series of possible reasons why audience members might consume a media text:

→ **Diversion** – escape from everyday problems and routine.
→ **Personal relationships** – using the media for emotional and other interaction, e.g. substituting soap operas for family life.
→ **Personal identity** – constructing their own identity from characters in media texts, and learning behaviour and values.
→ **Surveillance** – Information gathering, e.g. educational programmes, weather reports, financial news, holiday bargains.

Denis McQuail suggests a more detailed breakdown of audience motivation:

→ **Information**
  → finding out about relevant events and conditions
  → seeking advice
  → satisfying curiosity and general interest.
→ **Learning**
  → self-education
  → confidence through gaining knowledge.
→ **Personal identity**
  → finding reinforcement for personal values
  → finding models of behaviour
  → identifying with 'celebrities' – e.g. Beckham
  → gaining insight into oneself.
→ **Integration and social interaction**
  → gaining insight into circumstances of others
  → identifying with others – a sense of belonging
  → finding a basis for conversation and social interaction
  → having a substitute for real-life companionship
  → helping to carry out social roles
  → enabling one to connect with family, friends and society.
→ **Entertainment**
  → escaping, or being diverted, from problems
  → relaxing
  → getting intrinsic cultural or aesthetic enjoyment
  → filling time
  → emotional release
  → sexual arousal.

Action point

Think about all the media texts you have chosen in the last week – what were your reasons for choosing each of these texts?

Check the net

Remember that you can get viewing figures for televison programmes from BARB (www.barb.co.uk), for radio from RAJAR (www.rajar.co.uk) and for newspapers/from Media Guardian (http://media.guardian.co.uk), for example. These can allow you to do primary research about audience figures, not simply rely on these theories to write about how audiences consume texts.

Watching TV soap operas

A major focus for research into why and how people watch TV has been the genre of soap opera. Richard Kilborn suggests the following common reasons for watching soaps:

→ a regular routine and a reward for work
→ social and personal interaction ('Everyone's talking about it.')
→ fulfilling individual needs: a way of escaping or of enduring loneliness
→ identification with characters (perhaps cathartic)
→ escapist fantasy (especially supersoaps such as *Dallas*)
→ focus on topical, moral and social issues (e.g. in *EastEnders*)
→ appreciation of genre conventions.

A more recent model argues that the power lies with the individual, who will select the media texts that best suit their needs. The psychological basis for this model is the hierarchy of needs identified by Maslow. In other words, applying Maslow's hierarchy, because the audience member will have fulfilled the first four levels of the hierarchy (physiological needs, e.g. warmth and food; safety needs, e.g. shelter; belonging needs, e.g. family and friends; esteem needs, e.g. education), they can operate on the top level of self-actualisation and choose texts which directly meet their needs and desires – as indicated by Blumler and Katz.

Criticisms of uses and gratifications theory ●●●

Uses and gratifications can be a bit simplistic. It can be very hard to explain why we like something. It's also important to remember that TV viewing can be an end in itself. Sitting down at the end of the day and channel hopping might not be very active but can be just what the viewer wants to do!

A second problem is that we choose the media that we consume from what is available. The advert that you watch on the television may be extremely sexist. However, you have no control over what ads are shown on any one channel. This undermines the idea of **uses and gratifications** – we may not have enough control over the media products that we want. For example, many minority groups argue that they do not get the kinds of texts which they would like to see.

Other theorists, for instance David Morley, admit that individual differences do exist, but focus on the importance of socio-economic differences in shaping the ways in which people interpret texts (via shared **cultural codes**). In other words, this text will appeal to an A/B audience and another to a C/D audience. This again implies an active audience but is a bit heavy-handed in approach – which is what led to **reception theory** as a more sophisticated model to explain audience behaviour.

> **Checkpoint 2**
>
> What is a cultural code?

Exam question   answer: page 116

Look again at the TV listings which you analysed for the section on effects theory. What evidence of audience uses and gratifications in consuming media texts can you see from this analysis?   (1 hour)

Narrative theory

Narrative theory is one method of deconstructing the narrative of a text in detail – you should be aware of the methods of key narrative theorists and how they might be applied to a text, as this can often help you write about the **context of consumption** and audience responses to a text. There are four main theorists who are usually covered in an A level course: Propp, Todorov, Lévi-Strauss and Barthes.

Propp ●●●

Vladimir Propp published his *Morphology of the Folk-Tale* in 1928. He studied folk tales and legends from many different countries and noticed that they tended to be very similar. They seemed to be about the same basic problems and the same basic character types cropped up in most of the folk tales.

Based on this study, he identified 32 basic **categories of action**, which he called 'functions'. He also identified a set of basic **spheres of action** or character functions. He focused on the way that characters in folk tales tended to be types rather than individuals. There are eight of these character types that you should know:

- → the **hero**, who is on the quest (search)
- → the **villain**, who opposes the hero
- → the **donor**, who helps the hero by giving them a magic tool
- → the **dispatcher**, who starts the hero on his way
- → the **false hero**, who tempts the hero away from his quest
- → the **helper**, who helps the hero
- → the **princess**, who is the reward for the hero
- → her **father**, who rewards the hero for his efforts.

Of course, while there are many texts which do conform to these principles (*Star Wars* is a very well-known example), you will find at least as many texts which do not seem to do so. You might be able to think of examples of text where the **protagonist** (hero) or **antagonist** (villain) cannot be considered as types in this way (especially in indie or alternative films) – or perhaps a film where the '**princess**' does not conform to the function of a princess – *Alien* might be an example here. But just think about all those Hollywood films where the primary function of the heroine is to be the reward for the hero!

The important thing is that you can write about how Propp might or might not be a useful way of deconstructing a text which you are studying. It may be that talking about types will give you a lead in to write about representational issues, for example, or perhaps considering why Propp's theories are not appropriate when deconstructing a text will lead you to discuss the **ideology** of the text in some depth or to think in detail about the **audience**'s relationship with the text.

Claude Lévi-Strauss ●●●

Lévi-Strauss was a structuralist philosopher. What interested him was how much of our world is described in terms of opposites (or **binary**

Watch out!

The dictionary definition of **narrative** is 'giving an account of any occurrence or storytelling'. A word which often gets used in the same places as narrative is 'plot'.
In Media Studies, we usually use the word 'plot' to mean the events which occur in a text. Narrative usually refers to the structuring of these events – which is why we talk about narrative theory (the study of how texts are structured and ordered), not plot theory!

Checkpoint 1

How did Propp say that narratives are structured?

Action point

You may notice that the **functions** identified by Propp are very similar to the stages of classical tragedy defined by Aristotle and used by Shakespeare – i.e. exposition, rising action, climax, falling action, resolution. What does this suggest about the similarity of plots since people have been making plays and films?

Checkpoint 2

How did Lévi-Strauss analyse narratives?

oppositions) – night/day, good/bad, light/dark. He noticed that these oppositions tended to structure texts such as stories, plays and films as well.

This is easy to think about in terms of typical film but you can also analyse other texts in terms of these binary oppositions:

→ Washing powder adverts rely on 'before and after' contrast to convince you to buy the product.
→ News reports tend to be structured as binary opposites – the 'good' and the 'bad', because it means they can present the story very simply.

Rolande Barthes

Barthes was interested in concepts such as **negotiated meaning** between institution and audience. He argued that the reader produces new meanings when reading a text, making use of previous experience as well as the text itself. So, for example, when watching a docusoap about an airline, individuals' 'meaning' depends as much on their own experiences of airlines and docusoaps as on the text itself. The **cultural context** of consumption becomes as important as the **content** of the text. He described texts as '**networks**', which relates closely to current multimedia texts such as websites and the different ways in which a 'surfer' interacts with these non-linear texts.

The most well known of Barthes' codes is probably the **enigma code**. In its most basic form, it is the hook or mystery to be resolved for an audience, e.g. in serials which make use of the old 'come back next week to find out what happens' technique, or in trailers for new films.

Todorov

Tzvetan Todorov suggested that there are five stages to a narrative:

→ equilibrium
→ a disruption of this equilibrium by an event
→ a realisation that a disruption has happened
→ an attempt to repair the damage of the disruption
→ a restoration of equilibrium – which may be a new equilibrium.

This structure can be applied to fictional and non-fictional texts and can be a good starting point for identifying the structure of a text – just think how many news stories depend on establishing how an equilibrium is disrupted.

Adverts often use this structure to establish that the product being advertised is the solution to a problem. Think of washing powder adverts again and also think about food adverts, for example where the 'mother' has to create a 'quick but nutritious meal' for a hostile family – who instantly become a model family when confronted with this meal as solution to the disruption (i.e. hunger)!

Example

It has been suggested that one of the reasons why the media tended to refer to the 'war against Saddam' rather than the 'war against Iraq' was because of the vast amount of news to which we get access. Saddam was the binary opposition – a figurehead and 'demon' figure whom the audience could all hate.

Checkpoint 3

What is an enigma code?

Checkpoint 4

How did Todorov argue that a narrative is constructed?

Exam question   answer: pages 116–117

Watch a trailer for a new film (perhaps from www.imdb.com) and see if you can analyse the narrative of the trailer in terms of these four theories. Which seems the most useful for this trailer? Why?   (1 hour)

Representation and regulation

Checkpoint 1

What is a stereotype?

Action point

Many sketch shows use stereotypes in most of their sketches. Why do we find these funny?

It is important to be aware of key legislation relating to media representations and the regulatory bodies which monitor and control media production. The media have great influence in terms of how individuals and groups are represented and the legislation is there to ensure that individuals and groups are not misrepresented or abused in any way.

Stereotypes

A stereotype is a standardised representation of a specific group of people or objects. Stereotypes assign a limited number of characteristics to all members of a group. While we commonly use the term as it is applied to human beings, it is quite possible to stereotype objects as well.

Groups can be stereotyped by:

→ **age** ('All teenagers love rock and roll and have no respect for their elders.')
→ **sex** ('all women like housework.')
→ **race** ('All Eskimos look and think alike.')
→ **religion** ('All Catholics eat fish on Fridays.')
→ **vocation** ('All lawyers are corrupt.')
→ **nationality** ('All Germans are Nazis.').

Objects can be stereotyped around characteristics of:

→ **places** ('All cities are dangerous.' 'In England, it rains all the time.')
→ **things** ('All Skoda cars are cheap and badly made.')

Stereotypes are simplistic and unidimensional representations of a culture, or significant subgroup within that culture, which label a specific group of people. They are direct expressions of beliefs and values. A stereotype is a valuable tool in the analysis of popular culture because once it has been identified and defined, it reveals the ideologies which are operating within the text.

Basic characteristics of stereotypes

→ Simplistic: a stereotypical representation does not allow for complexity or depth.
→ Secondhand: people absorb stereotypes from their cultural **context** rather than from experience with members of the groups being stereotyped.
→ Sometimes false: they are attempts to claim that each member of a group shares a set of common qualities, which may not be true.
→ Stereotypes rarely change over time because social and cultural attitudes change very slowly – but they can be deliberately subverted in a media text.

Media regulation

There is a lot of regulation to ensure fair representation in the media although this does not always stop bias or complaints.

→ **The Race Relations Act (1976)** states that media institutions may not broadcast or publish any material which might be deemed offensive to ethnic and racial groups.

→ **The Obscene Publications Act (1959, redrafted 1964)** ensures that nothing obscene could be represented in the media. So, for example, the 9 p.m. 'watershed' remains important for broadcasters, not just in terms of content of programmes shown pre-watershed, but also in the representations that may be used.

→ **The Representation of the People Act (1983)** sets out guidelines about how much broadcast time is allowed for each of the main political parties during a year and how they may use it.

→ **The Public Order Act (1986)** prevents the broadcast or publication of any material which could be considered offensive to any ethnic or national group.

→ **The Sex Discrimination Act (1986)** ensures equality of representation for both genders in the media and in industry.

→ **The Broadcasting Act (1990)** states that the ITC (Independent Television Commission) must ensure that programmes broadcast will not be offensive in any way.

Regulatory bodies

Most regulatory bodies for the media are voluntary – broadcasters, print organisations and other media producers do not have to subscribe to the relevant groups, although most do. It is only the BBC, being a Public Service Broadcaster, which has these regulations written into its charter. These bodies include:

→ **Advertising Standards Agency**, which is responsible for maintaining standards in adverts and addressing complaints from members of the public who find an advert offensive in some way.

→ **Press Complaints Commission**, which fulfils the same role in relation to material published in newspapers and magazines.

→ **OFCOM**, which is now the major regulator for all the communications industries, responsible for broadcasters, phone companies, wireless communications and so on. It replaced the regulatory bodies which were separately responsible for each of these industry sectors and is not a self-regulatory body.

Action point

What can you do if you think these laws have been breached in a media text? You might want to debate this in class.

Examiner's secrets

Although you need to know what these laws are, you do not need to be able to write about them in depth or offer case studies about them in your answers.

Checkpoint 2

What is a self-regulatory regulatory body?

Exam question   answer: page 117

Watch an episode of a popular sitcom and an episode of a popular soap opera. Which one makes more use of stereotypes? Why is this?   (1 hour)

News values

Checkpoint 1

How would you prioritise these news values?

Galtang and Ruge outlined the first list of news values in 1965. They suggested that the selection of news for newspapers and for news broadcasts is done by application of these news values – explicitly or implicitly – by the 'gatekeepers' who control the news we receive.

Galtung and Ruge ●●●

The values they identified are:

→ **Frequency:** The time-span of an event and the extent to which it fits the frequency of the newspaper's or news broadcast's schedule.
→ **Threshold:** How big is an event? Is it big enough to make it into the news?
→ **Unambiguity:** How clear is the meaning of an event?
→ **Meaningfulness:** How meaningful will the event appear to the receivers of the news? Hartley stresses in this context what he refers to as 'cultural proximity'. Events happening in cultures very different from our own will not be seen as being inherently meaningful to audiences here.
→ **Consonance:** Does the event match the audience's expectations? Journalists have a pretty good idea of the angle from which they want to report an event, even before they get there.
→ **Unexpectedness:** 'Man bites dog' is news. If an event is highly unpredictable, then it is likely to make it into the news.
→ **Continuity:** Once an event has been covered, it is convenient to continue to cover it – the running story.
→ **Composition:** This is a matter of the balance of the news. It is a matter of the editors' judgement, more than anything else. A different news broadcast will have a different agenda in terms of 'hard' and 'soft' news, for example, usually dependent on the perceived target audience.
→ **Reference to élite nations:** This relates again to 'cultural proximity'. Those nations which are culturally closest to our own will receive most of the coverage.
→ **Reference to élite persons:** The media pay attention to important people. Anyone the media pay attention to must be important (this links to cult of the celebrity).
→ **Personalisation:** This connects with unambiguity and meaningfulness. Events are seen as the actions of individuals.
→ **Negativity:** Bad news is good news in terms of what is reported.

Action point

Do you think that the availability of good footage influences the shape of a news bulletin?

You will probably feel that there are certainly some news values missing from the list – available resources, for example. After all, a TV news report with no relevant images would probably not be used without very good reason. Nor does the list mention entertainment – yet most TV news bulletins will seek to finish with an entertaining story and this may well take precedence over a hard news story.

Schultz

Shultz analysed news values in a different way, categorising them into six **dimensions**:

→ **status** (elite nations, persons and institutions)
→ **identification** (proximity, ethnocentrism, personalisation, emotion)
→ **valence** (aggression, controversy, values and success)
→ **consonance** (predictability, running themes/issues, stereotypes)
→ **relevance** (consequence or cultural importance, etc)
→ **dynamics** (unexpectedness, timeliness, etc).

Gatekeeping

The gatekeepers are primarily the editors, who are responsible for selecting and ordering the news broadcasts or selecting and planning the newspaper layout. The gatekeeper will only allow particular items into the news text and this selection will depend on factors such as the application of news values, audience expectations and institutional requirements. In other words, the selection of news events is not a reflex action, but the **socially determined construction of reality**.

There are other, more complex models to represent the gatekeeping process but this is a good basic illustration.

Bias in the news

At one time or other we all complain about bias in the news. The fact is, despite the journalistic ideal of objectivity, every news story is influenced by the attitudes and background of its interviewers, writers, photographers and editors – even before the gatekeeping process! Among the primary forms of bias which may affect the news text are:

→ bias through selection and omission
→ bias through placement
→ bias by photos, captions and camera angles
→ bias through use of names and titles
→ bias through statistics and crowd counts
→ bias by source control
→ word choice and tone.

When you are evaluating a news text, therefore, you need to be aware of news values employed during selection and construction; how the text has been mediated by gatekeeping and what bias may be operating. In other words, the ideologies behind a news text can be very powerful but few news texts make them explicit. As you have seen, tabloid newspapers often do make these explicit – BBC News bulletins may not. When deconstructing these texts, you must consider these factors.

Checkpoint 2

What is gatekeeping?

Check the net

If you look at the Sky News website and visit the education section they have some useful information about gatekeeping and the editorial process at Sky News.

Checkpoint 3

What do we mean by bias?

Exam question   answer: page 117

Record an evening news broadcast. Analyse this broadcast in terms of news values employed during selection and construction, possible gatekeeping and possible bias in the broadcast. How explicit are these factors for the audience? Does this matter?   (1 hour)

Audience profiling

The relationship between text, institution and audience is an important one and media organisations do not create media texts without clearly defining a target audience. Advertisers, in particular, are keen to define target audiences very carefully so they can establish the right ideology to make the target audience want the product and so they know where to focus the campaign. There are many ways of profiling audiences but the basic forms of demographic and psychographic profiling are mentioned here.

Checkpoint 1

What is demographic profiling?

Links

Maslow's hierarchy of needs is shown on p. 91.

Checkpoint 2

What is psychographic profiling?

Action point

Which of these groupings do you feel you belong in? Is this typical of other members of the class?

Demographic profiling

→ The most basic form of identification for target audiences is demographic profiling.

→ This way of grouping an audience defines them in terms of age, class, gender, geographical area, class, economic status and religion.

→ This is a very simplistic way of defining an audience because it assumes that everyone in a very broad group has the same attitudes.

Psychographic profiling

To move away from these very broad audience groups that are difficult to target, market researchers started to categorise audiences in terms of needs and motivation rather than simple demographic factors. In this way consumers can be targeted directly through their needs and desires, which is obviously more effective for advertisers – and for broadcasters and publishers.

Advertisers, in particular, aim now to link the ideologies of the product to those of the consumer – so as to generate better empathy and bonding with the brand. The emotional equity or loyalty a brand holds today is gaining importance as the market gets cluttered with brands which are not well differentiated in terms of physical attributes (e.g. different brands of coffee).

The four Cs

The four Cs (Cross-Cultural Consumer Characteristics) categorise people in terms of their personal **aspirations**. This is one of the earliest but still most popular ways of profiling audiences since it profiles the audience in terms of wants and needs, not simple demographics.

→ **Mainstreamers** – This is the largest group. They are concerned with stability and security, mainly buying well-recognised brands and consuming mainstream texts.

→ **Aspirers** – Aspirers are seeking to improve themselves. They tend to define themselves by the high status brand names which they own and consume, absorbing the ideologies of the products as their own and believing that their status is established by this conspicuous consumption.

→ **Succeeders** – These are people who feel secure and in control – generally, they are in positions of power, although they may not be. They buy brands which reinforce their feelings of control and power.

→ **Reformers** are idealists who actively consume eco-friendly products and buy brands which are environmentally supportive and healthy. They buy products which establish this 'caring and responsible' ideology.

→ The **individual** is an addition to these four basic groups – the individual is highly media-literate, expects high-production advertising and buys product image, not product, requiring high-profile, sophisticated campaigns.

Lifestyle marketing

Media organisations sometimes seek to 'nickname' particular niche markets as an easy way to define target audiences. Among the best-known names are: YUPPIES (young upwardly mobile professionals), DINKYS (double income no kids yet).

Some of the latest analysis is done using particular tools for market research, such as LifeMatrix, the psychographic marketing tool launched in December 2002 by MRI and RoperASW. This tool defines ten categories which are much more centred around values, attitudes and beliefs than the original, more simplistic categorisations.

Checkpoint 3

Why do you think advertisers create these niche nicknames?

The LifeMatrix segments

→ **Tribe Wired**: Digital, free-spirited, creative young singles.
→ **Fun/Atics**: Aspirational, fun-seeking, active young people.
→ **Dynamic Duos**: Hard-driving, high-involvement couples.
→ **Priority Parents**: Family values, activities, media strongly dominate.
→ **Home Soldiers**: Home-centric, family-oriented, materially ambitious.
→ **Renaissance Women**: Active, caring, affluent, influential mums.
→ **Rugged Traditionalists**: Traditional male values, love of outdoors.
→ **Struggling Singles**: High aspirations, low economic status.
→ **Settled Elders**: Devout, older, sedentary lifestyles.
→ **Free Birds**: Vital, active altruistic seniors.

Check the net

You can find out more about LifeMatrix at http://www.roperasw.com/products/lifematrix.pdf

The uses of audience profiling

Of course it is important, as with all theories, not to be too focused on audience profiling. Remember that is primarily only advertising organisations who will complete elaborate market research to profile an audience in great depth because this helps them to construct a specific ideology and image for a product directly relating to the attitudes, values and beliefs of that target group.

However, knowing about the different ways of profiling audience groups and being aware of the limitations of demographic profiling is important. If you are able to research and study your texts with a greater understanding of the target audience, you will be able to look at your texts in a much more appropriate way, avoiding simplistic statements like: 'My film will appeal to a female audience'.

Exam question   answer: page 117

Choosing a current advertising campaign, profile the target audience using some of the groupings here. In what ways do these categories relate to the ideologies established by the campaign and the implicit values, attitudes and beliefs of this audience?   (1 hour)

Advertising

Links

You should look at the section on audience profiling on pp. 98–99 to get a clearer idea of how advertising agencies profile their target audiences in order to sell to them.

Checkpoint 1

What is intertextuality?

Checkpoint 2

What is product placement?

It is important, as well as knowing about the advertising form, to have an understanding of the advertising industry, types of advertising and the construction of brand image. Advertising campaigns are expensively constructed and marketed and rely on high production values to establish a high-value image for the product. It is also important to sell the right product to the right audience so advertisers take care to display/show their campaign in places which will generate maximum exposure to their primary target audience. Of course, they may also need to communicate to a secondary audience – as with the toy adverts usually shown at Christmas!

Analysing an advertising campaign

There are always some key starting points to think about when researching advertising campaigns:

→ Is the campaign based on a single, high production value advert with maximum exposure, a single theme across a range of media or a series of adverts – or even a narrative?
→ What factors enable you to profile the target audience?
→ What contexts has the campaign appeared in? How do these relate to the ideology of the advertising campaign?
→ What ideologies are established for the product? How do these relate to the audience profile?
→ What cultural assumptions are made by this campaign? Does it make use of intertextuality to establish credibility or context?

Covert advertising

Remember that not all advertising is overt, or obvious. **Covert** advertising can range from product testing sections in magazines through **product placement** where certain products or adverts can be seen in shot during a programme (think of advertising during football matches), to even more covert techniques such as **sponsorship** of programmes or paying a celebrity to be seen using a particular product. All these techniques have the same objective – to promote the product and make the audience aware of it and aspire to having the product. Once they aspire to the product they are likely to buy it. One recent form of covert advertising is **viral advertising** where adverts are created to be sent between friends' mobile phones, for example, promoting a product without needing to pay for a slot or produce a poster.

Product placement

TV and film may well make deals with advertisers to promote their products by using them in their films. The deal between *The Matrix* and Nokia was a famous example of this. The advantages are:

- There is a captive audience and wide exposure (especially with high-profile films and TV programmes).
- It is cheaper than direct advertising or sponsorship.
- Ideologies of product and context become linked, which strengthens the product image for the target audience.

Sponsorship

Sponsorship is now common on UK TV, following the US model. So, *Coronation Street* is sponsored by Cadbury and *Friends* by Jacob's Creek, for example. In this way, again, the advertiser can link the ideologies of the product with the programme and create a stronger product image for the target audience. This is why sponsorship is considered covert advertising, since the sponsors do not need to establish ideologies for the product but borrow those of the programme being sponsored.

Plugs

- At the most basic level, plugs can be the free trial stands in a supermarket where you can taste the product being promoted.
- Promotional offers are also a form of plug. Film companies will offer free promotional gifts for customers buying products at a fast food outlet, for example – again linking the ideologies and getting greater exposure for the product. (This can work very well as it gets exposure with children who might not otherwise know about the film – bringing 'pester power' into play.)
- The best-known form of plugging is, of course, celebrities appearing on TV chat shows to promote their latest film/book/TV series/yoga video, etc). There will often be a big campaign, with lots of appearances by stars, magazine interviews, etc to plug a new blockbuster film, for example.

Viral advertising

Viral advertising first appeared when someone produced a parody of a well-known advertisement and sent it as an email attachment to some friends. The email was passed on and spread quickly, like a virus, rapidly creating a lot of publicity for the product. Now some advertising agencies are developing deliberate viral campaigns to promote their products, e.g. for alcohol and clothing. It is very cheap advertising and is an ideal way to reach younger target audiences who frequently share emails and mobile phone messages in this way. These are popular audiences to target, of course, because they are seen as having high disposable income.

Example

Disney will often promote a new children's film by producing a range of merchandise such as figures of characters from the film, which are available at McDonald's, for example. McDonald's are happy as children want to go in order to get the figures and Disney are happy because it promotes the film to the primary audience (children).

Check the net

You can find a selection of viral adverts and email attachments at www.virall.lycos.co.uk but be warned that some of them might not be very tasteful!

Check the net

Most of the well-known advertising agencies promote their work on the internet. For example; M&C Saatchi, Saatchi & Saatchi, McCann-Erikson, Ogilvy & Mather, Abbott Mead Vickers BBDO.

Exam question   answer: pages 117–118

Use the Internet to explore one of the main advertising agencies in the UK. Choose a current campaign and research the campaign in terms of target audience, audience profile, context of display, ideologies established and what image of the product is created.   (1 hour)

Postmodernism

Examiner's secrets

Don't think that you need to be able to 'define' postmodernism in your essays – by definition it is undefinable!

Postmodernism is a difficult concept to define – perhaps because it is intended as a rejection of conventional reality and conventional forms and therefore it tends to be defined in terms of what it is not.

Postmodernism refers to ideas after modernism. So postmodernism identifies contemporary culture and society, which has developed since modernism. The term is complex and hotly debated so we are not going to attempt to define it here. This section will introduce you to some of the important concepts within postmodernism which you will find useful when writing about postmodernist media texts.

Postmodernism deliberately rejects the boundaries between high and low forms of art and rigid genre distinctions, emphasising instead pastiche, parody, bricolage, irony and playfulness. Postmodern texts favour reflexivity and self-consciousness, fragmentation and discontinuity (especially in narrative structures), ambiguity, simultaneity and an emphasis on the destructured, decentred, dehumanised subject.

Intertextuality and bricolage

Intertextuality and **bricolage** are important concepts in relation to postmodernism. Intertextuality is the way that one text may make use of a different text, in order to add a further layer of meaning. So, for example, an advert might be intertextual if you need to have watched a particular television series to know what is happening in the advert. Bricolage takes this a stage further. Bricolage suggests that this referentiality has been used in such a way that a new meaning has been constructed. For example, the fusion of documentary, game show and soap opera which became reality TV was a form of bricolage, giving a new meaning and structure from this combination.

Links

You can see that the concept of media 'consumption' is part of Jameson's third stage of development.

Jameson

Jameson identified stages of capitalism which have led to the development of the postmodernist culture and society:

1. **Market capitalism**, associated with particular technological developments, i.e. the steam-driven motor, and with a particular kind of aesthetics – realism.
2. **Monopoly capitalism**, associated with technological developments, especially electricity – and modernism in art.
3. **Multinational or consumer capitalism**, associated with sophisticated nuclear and electronic technologies, and correlated with postmodernism.

His focus was on the commodification of culture as economic pressures begin to shape culture directly. For example, early film makers were able to make films which interested them. Today's media producers must satisfy the investors first.

Lyotard

Lyotard argues that society maintains stability through metanarratives, which are stories a culture uses to structure its practices and beliefs

(e.g. religion, political belief). Every belief system or ideology has its metanarratives, according to Lyotard – for instance, in Marxism, the metanarrative is that capitalism will collapse and a socialist world will evolve. He argues that in a postmodern world these metanarratives are no longer central and therefore our society is no longer stable, but fluid. For Lyotard then, postmodernism is the understanding that these metanarratives mask the contradictions and instabilities that are inherent in any culture. Postmodernism constructs 'mini-narratives' instead – stories that explain small practices or local events, rather than large-scale universal or global concepts.

Baudrillard ●●●

Baudrillard is probably most famous for his statement that the 'Gulf War never happened'. This statement makes more sense (in a postmodern way!) than it might seem. For Baudrillard, a postmodern society is comprised of simulacra, not originals – not only the obvious simulacrum of a poster copy of a famous painting but also a television programme, for example – there is no original programme which all the others copy, only the copies themselves. Another illustration of Baudrillard's simulacrum would be computer games, e.g. *Sim City* or *Virtual Zoo*. These are a reality created as a simulcrum which cannot (by definition) have an original.

Baudrillard calls this simulation a 'hyperreality' and it is this hyperreality which led to the statement about the Gulf War. He was referring to the way we receive news 'live' from around the globe, experiencing the war as a simulation in our own cultural context. Due to this simulation we react as though what we are experiencing is real. Baudrillard was therefore referring to the way that the media presented a 'reality' of the war which actually bore little resemblance to the reality of the war itself. The simulacra (the media texts) have become more 'real' than the reality they seek to convey.

Simulacra can be seen throughout contemporary society, for example:

→ Computers using a GUI (Graphical User Interface), e.g. Windows to hide the code which is working inside the machine.
→ Disneyland and other theme parks.
→ Flight simulation games (and indeed real flight simulators used to train airline pilots!).
→ MUDs (Multi User Dungeons) for online gaming – and MOOs (MUDs, Object-Orientated), their more sophisticated versions.
→ Virtual communities, e.g. peer-to-peer networks.
→ 'Live' presentations of events which are still constructed (whether by time lapse in case someone says something unfortunate, or simply by a range of cameras being used and the director being able to construct a reality by selecting different cameras.
→ 'Live' news or sport, especially if an event is edited down to 30 seconds of highlights to be transmitted to a mobile phone.

Checkpoint 2
What is hyperreality?

Action point
How many simulacra do you meet in some way during a day? How much of your view of reality is based on these simulacra?

Example
You probably have a fairly good idea about what New York looks like, even if you have never been there – because you have seen simulacra of it so often.

Exam question
answer: page 118

To what extent would you say that contemporary television news broadcasting is postmodern?   (1 hour)

Genre theory

Checkpoint 1

What do we mean by media-saturated?

In earlier sections you have focused on genre as a key concept for Media Studies, helping you to analyse texts and categorise them as you study. You have also identified the principles of generic codes and conventions in texts as a way of identifying the genre and fulfilling audience expectations for texts in that genre. In this section we look at genre theory, which moves beyond this to identify how genre is culturally constructed and the relationship between genre and cultural ideologies.

Defining genre ●●●

John Fiske describes genres as 'attempts to structure some order into the wide range of texts and meanings that circulate in our culture for the convenience of both producers and audiences'. Genre is, therefore, simply a way of categorising texts in our media-saturated culture.

Genre is:

→ a way of categorising texts
→ a way of defining codes and conventions for a category of text
→ a way of setting up audience expectations
→ different from form – form means the shape of a particular text, genre refers to a type of text
→ culturally dependent – different cultural groups construct different genre expecations
→ constructed through a series of signs (visual and aural) which are associated with that genre or through use of generic narratives and ideologies.

Links

If you want to find out more about these hybrid forms of documentary you can look at p. 66.

Culture and genre ●●●

Genre is a **dynamic** concept, which changes over time and in different cultural contexts. Of course, because these changes are gradual and culturally dependent you may not realise that the change has taken place. The changes may be due to wider issues such as desensitisation or due to institutional developments – such as when **hybrid** genres start to appear, which then become genres in their own right after a period of time. **Docusoap** was initially considered very much a subset of documentary, and **reality TV** as a combination of cinema verité, soap opera and game show is a fairly recent genre development.

Cross-cultural expectations

It is easier to be aware of generic codes and conventions when looking at genres with which you are less familiar. For example, many Western audiences are able to identify clearly the generic elements of a typical Bollywood musical, even on first viewing. Of course, this awareness can also be used by media producers to their advantage. On the most basic of levels, think of the number of television programmes which simply show clips of advertising or television programmes from other countries – to the accompaniment of a laughing studio audience!

Background to genre theory ●●●

Genre theory is generally considered to have developed as a result of André Basin's work on Westerns in the 1950s. He was the first to explore the very significant advantages of genre from an institutional perspective.

Genres make film making more efficient (by allowing the re-use of plots, sets and the like) as well as more marketable (by using generic conventions as a way of 'selling' the film to the audience). Genres in film, therefore, were seen as more than arbitrary: they originated at the level of production.

Altman argues that genres are usually defined in terms either of certain media language (taking the Western as an example: the guns, horses, wagons, towns, landscapes, or even the Western stars such as John Wayne or Clint Eastwood) or certain ideologies and narratives. He calls the former group the *semantic* elements and the latter the *syntactic*, and argues that genre theory needs the two to be kept separate as a means of dealing with genre progress over time and hybridity.

Questions to ask when exploring genre theory across a series of texts ●●●

→ What is the evolutionary history of the genre?
→ What is the relationship between particular texts and their genre?
→ What genre(s) are operating in these texts? Are there some hybridity or cross-genre elements here?
→ How do texts compare across the history of a genre?
→ How do texts compare culturally across this genre?
→ What semantics are employed within this genre?
→ What ideologies are established at the syntactic level in this genre?
→ What narratives are established at the syntactic level in this genre?
→ What factors have influenced the success of this genre over time?
→ Is this a cross-media genre? If so, how is it mediated by the different medium?
→ What technological factors have affected the development and progression of this genre?
→ How are audience expectations established/maintained within this genre?
→ What institutional advantages does this genre offer for media producers?

Checkpoint 2

What is the relationship between the semantic and the syntactic elements of a media text?

Don't forget

Genre is a flexible concept and culturally dependent so genre expectations and codes and conventions can vary.

Exam question   answer: page 118

'Audiences feel safe with the idea of genre.' How far do you feel this statement to be true?   (1 hour)

Star theory and auteurs

Institutions make use of stars and auteurs to construct an audience for a text. However, it is important to realise that these ideologies are, in fact, constructions, intended to persuade audiences to engage with the texts.

Stars and the Hollywood studios

The early Hollywood studio system was, in many ways, like a factory. The primary objective of the studios was to make money and stars became a useful commodity to do that. During the Golden Age of the Hollywood system (1930s), the moguls, or studio heads, had the power and the resources to produce an incredible number of films.

The rise of the studio system depended on the stars, who were constructed and exploited to suit a studio's image and schedule. Actors and actresses were contract players bound up in long contracts with a single studio, and had little choice about what films they would appear in. Stars could be loaned out to other production companies at any time. Studios could also force bad roles on actors, and control the stars' images with their in-house publicity departments.

Star image

The concept of the star is an image. The star persona is constructed in line with the ideologies of the institution and is intended to appeal to the audience. This image often has little relation to the reality of the star.

Dyer has identified key codes and conventions of the star persona:

→ A star is a constructed image, constructed across a range of media and mediums.
→ Stars are **commodities** produced by institutions.
→ Stars represent particular ideologies for us and are available for us to explore these ideologies – for example, through articles in newspapers or the latest gossip.
→ Stardom, and star worship in general is a cultural value in itself, often related to both materialism and ideologies of beauty and sexuality.
→ Stars support **hegemony** by conforming to it (thin/beautiful) or providing difference (fat/still lovable). Much of the discussion of stars in celebrity magazines is about how stars compare to the **hegemonic ideal**, and how we compare to the stars.
→ Film stars are represented primarily through their roles – constructed by screenwriters. The screenwriters create the personality and characteristics for them, and their overall image is constructed from many fragmented parts, which may or may not contradict each other.
→ Stars are generally **aspirational** for an audience – hence why their hairstyles or clothes can be so successful. We feel that if we buy into their culture materially we will somehow also be perceived in the same way.

Checkpoint 1

What is a media mogul?

Don't forget

A star is not the same as a person. It is an image or representation constructed for a particular context (e.g. Hollywood) and may well bear very little relation to the reality of the person behind the star.

Auteur theory ●●●

A **film director** directs the artistic and dramatic aspects of a film by:

→ defining the artistic vision of the film
→ controlling the content and flow of the film
→ directing the actors
→ organising locations in which the film will be shot
→ managing technical aspects such as the positioning of cameras, the use of lighting and the timing and content of the film's soundtrack.

In practice the director will delegate many of these responsibilities, e.g. by using the storyboard to visualise a scene and then leaving it to the crew to find a suitable location, or to set up the appropriate lighting.

The degree of control that a director exerts over a film varies greatly. Many directors are essentially employees of the institution, which may have particular views about the film under construction. This was especially true during the Golden Age when directors had little freedom – much like the stars they were directing.

In some cases, however, the director makes a very significant contribution to the form and ideologies of a film under construction and this form of director is known as an auteur. This term was first applied to French film directors by François Truffaut, who used the phrase 'la politique des auteurs' to describe an ideal vision of how a director should strive to prevent films simply being institutional vehicles. He argued that an auteur creates a film which has integrity and challenging ideologies. His use of the word 'politique' was apt – there should, for him, be clear ideologies explored within a film and communicated to an audience – not the kind of entertainment-driven film which the studios prefer.

Truffaut argued that art cannot be constructed by some calculated process – finding an 'important' subject, hiring a 'distinguished' playwright, finding 'authentic' locations and so on. A movie might fail in many ways and still be important as a revelation of what some creator thinks and feels if it has integrity in its own right and seeks to create meaning for an audience.

To identify a director as an auteur therefore is to imply certain expectations for an audience in terms of their films:

→ integrity
→ creativity
→ ideologicial questions and debates
→ artistic exploration
→ above all else – film making which seeks to communicate with an audience, not which is made to generate money.

Checkpoint 2

What is an auteur?

"There are no good and bad movies, only good and bad directors."

Truffaut

Action point

Using www.imdb.com as your starting point, research the films of a director you are interested in. Do their films match these expectations?

Exam question   answer: page 118

Choosing one director who you are familiar with, explain why you do or do not consider them to be an auteur.   (1 hour)

Gender studies

This section covers some of the most important theories related to gender or gendered studies. The theories of Laura Mulvey, for example, are very important, especially in relation to Hollywood film. Feminist theorists and queer theory provide different ways of receiving and reading a range of media texts.

Feminist theory ●●●

→ Feminism is the response to society's assumptions that women should be subservient to men. Until the emergence of feminism, women were treated almost as objects, passive agents in a male world.

→ The roots of the feminist movement extend back to the eighteenth century but run through the Suffragette movement which fought for the vote for women in the early part of the twentieth century and the land girls and other women who moved into many male occupations during the Second World War, for example.

→ Activists in the early 1970s were battling for Womens' Liberation and equal opportunites at a time of great unrest and social upheaval for many social groups. The Sex Equality Act was not passed until 1975.

→ Academic feminist theory has developed from this liberation and activism. This has led to a range of critical writing about the role of women in contemporary society and therefore also the role of women in contemporary media. Many theorists have debated the way the media construct representations of women and the ideologies they maintain. Among these is Laura Mulvey's theory of the **male gaze** as constructing audience expectations of film.

→ Patriarchy is a form of society ruled by men through the figure of the father (the patriarch) to whom all others are subordinate. A patriarchial society therefore is one where the men dominate and the media are constructed for them as audience.

→ Gender refers to those aspects of our representation which establish whether representations in a text are 'male' or 'female' in representation, narrative function, ideological function and so on.

The male gaze ●●●

Laura Mulvey argues that cinema audiences look at films in two ways – **voyeuristically** and **fetishistically**. Cinema audiences watch a film without being watched by the characters on screen and usually in a darkened cinema so other audience members do not observe them either. Therefore they are almost **voyeurs**, watching the people on screen. This can lead to two effects:

→ **objectification** of female characters in relation to this controlling (male) gaze

→ **narcissistic identification** with an ideal image seen on the screen.

She argues that this **voyeurism** involves turning the represented figure itself into a **fetish** (object) so that it becomes increasingly beautiful but more objectified. Fetishistic looking, she suggests, leads to the cult of

Action point

Using a range of texts which you know well, think about the gender representations employed. Do they sustain gender stereotypes?

Checkpoint 1

What is the male gaze?

the female movie star, celebrated for her looks but considered as an object and often treated as such.

Conventional Hollywood films have a male protagonist in the narrative and assume a male audience. Male characters are active and dynamic and not always conventionally attractive. Actresses, on the other hand, must be glamorous and attractive but are there only in supporting roles (the Princess according to Propp) and therefore are on screen as 'eye candy' to appease the **male gaze** of the (male) audience.

Gender as performance ●●●

Judith Butler argues that we all put on a gender performance, whether traditional (heterosexual) or not. By choosing to be different about it, we might work to change gender norms and the binary understanding of masculinity and femininity as represented in the media. She suggests that:

→ Gender is not some inner truth but the presence of received meanings.
→ Gender is not fixed but constructed.

If you think about the way that gender representations are used to construct identity in the media you can see how these 'performances' can be manipulated. For example, think of music artists who often construct complex gender ideologies as part of their image – Madonna has constructed images which are expressly feminine, overtly masculine, and less extreme over her career. There are many artists who seek to construct particular gender identities in order to present particular ideologies to an audience – Eddie Izzard and David Beckham both use feminine gender identity at times as part of their image, presenting less conventional ideologies to their audience. In order to change represent-ations and expectations, Sigourney Weaver shaved her head completely and was filmed in a far more 'male' context than had previously been the case for female characters in conventional sci-fi film, who, whether human or alien, were usually products of the male gaze.

Gender and lifestyle ●●●

David Gauntlett argues that in contemporary society, gender roles are more complex and that the media reflect this. He points out that female role models today are often glamorous as well as successful (Ally McBeal) in a way that previously they were not. He argues that much of this is due to the rise of **'girl power'** in the media, through identities constructed by music artists such as Destiny's Child as well as contemporary actresses, for example, who are demanding less passive film roles. He argues that our expectations of gender are flexible and culturally dependent and therefore will continue to change. He tracks these changes through different media texts, especially lifestyle magazines.

Checkpoint 2

What do we mean by 'eye candy'?

Action point

Why do you think Madonna plays with different forms of gendered representations in these ways?

Action point

Do you think gender expectations in media texts are changing?

Exam question   answer: pages 118–119

'The male gaze dominates Hollywood movies.' Using the top five films of the week, would you agree with this statement? Give reasons for your answer.

(1 hour)

Marxism

Marxism is a political practice and social theory based on the works of Karl Marx, a nineteenth-century philosopher, economist, journalist and revolutionary. Marx drew on Hegel's philosophy, the political economy of Adam Smith, Ricardian economics and nineteenth-century French socialism to develop a critique of society which he claimed was both scientific and revolutionary. This critique achieved its most systematic (if unfinished) expression in his masterpiece, *Capital: A Critique of Political Economy* (*Das Kapital*).

Marxist ideologies

Marx traced the development of material life, predicting that as capitalism replaced feudalism, so socialism would supersede capitalism.

Marx argued that capitalism is based on the profit motive and that under capitalism, profits are generated by exploiting workers (i.e. the value of sales for a product is greater than the cost of making it and paying the workers).

Marx believed that capitalist society is divided into two social classes:

→ the working class, or proletariat, who sell their labour and do not own the means of production
→ the bourgeoisie, who own the means of production and employ the proletariat. The bourgeoisie may be further subdivided:
  → wealthy bourgeoisie who do not also need to work themselves
  → petty bourgeoisie who employ others, but also work themselves.

Marxism and the media

→ Media producers produce media texts within this society, which maintain these social divides. Even when media producers attempt to produce texts which criticise this position they are doing so from within the society and therefore, by implication, accepting it since they are still working within it.
→ The **'culture industries'** (as Adorno described them) constantly seek greater audiences because of the profit motive. Therefore they will construct texts intended to generate mass audiences, hence dumbing down their output.
→ Mass media are seen as a way of entertaining the workers while drip feeding them ideologies and beliefs (effects theory).

Althusser

Althusser identified what he called the Ideological State Apparatuses (ISAs), which operate alongside the Repressive State Apparatuses (military, police, etc) to maintain these bourgeois ideologies:

→ religion
→ education
→ family (family values)
→ legal and political system
→ political system (including trade unions)
→ cultural and communications systems.

Links

Marxism is closely associated with effects theory so you may want to look at the section on effects theory on pp. 86–87.

Checkpoint 1

What is the difference between the proletariat and the bourgeoisie?

Checkpoint 2

What is the difference between the ISAs and the RSAs?

He studied the way these ISAs act to **interpellate** an audience (i.e. position the audience in such a way that they presume that these ideologies are correct).

Gramsci

Gramsci extended this with the concept of **hegemony** – the dominance of the ideologies of the ruling class, with all the beliefs and values that they incorporate. He argued that this process is so successful that it becomes 'common sense' and these ideologies become taken for granted. So, by maintaining these ideologies in the media, the ruling classes can ensure that they retain hegemonic control.

Eisenstein and montage

Sergei Eisenstein and many other Soviet film makers in the 1920s used Marxism as justification for their film making. They took a more structuralist approach, based on their Marxist beliefs, complaining that Hollywood cinema is designed to draw the audience into believing in the capitalist propaganda and ideologies. They felt that even conventional narrative structure and media language maintained these capitalist ideologies (e.g. camera angles being used to give power to important characters and narrative resolutions where the action hero wins the prize of the princess).

Eisenstein's solution was to shun narrative structure by eliminating the individual protagonist and tell stories where the action is moved by the group, so the audience cannot engage emotionally with the 'story'. In his films the narrative is told through **montage** (i.e. a clash of one image against the next, whether in composition, motion, or idea) so that the audience is never lulled into believing that they are watching something which is real.

Williams

Williams, a more contemporary Marxist, rejected the term 'mass culture' because of the implication that the products of the culture industries are somehow 'low culture', deemed suitable for workers, and have less status than 'high' culture, e.g. ballet and opera.

This concept of high and low (popular) culture still occurs today. High culture events are given a status very different from popular culture events, despite significant differences in size of audience. Culturally, we still differentiate between theatre and film, for example, with many US film actors still striving to appear on Broadway or in London theatre productions to prove that they are serious artists. Popular culture is still seen as second-class. This differentiation underlies many analyses of the differing positions of PSB and commercial broadcasting.

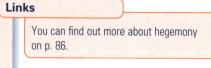

Links

You can find out more about hegemony on p. 86.

Checkpoint 3

What is montage editing?

Check the net

There is a good collection of links to information about Williams and his exploration of high/popular culture at http://www.popcultures.com/theorists/williams.html

Exam question   answer: page 119

Watch a film by Sergei Eisenstein, such as *Battleship Potemkin*. In what ways does he seek to use editing to create meaning for an audience? How does this compare to conventional Hollywood editing?   (1 hour)

Distribution and exhibition

Once a media text has been made, it must be distributed and marketed. For most media texts, the market will have been established before the text is made (e.g. a television programme is commissioned for a particular slot, on a particular station, and a magazine article for a specific edition). Films, however, have to be marketed more explicitly to generate the audience. Note that there is a very big difference at times between a film which is critically successful (lots of critics like it) and a film which is commercially successful (lots of people pay to see it). Of course, most blockbuster films should achieve both types of success but many do not. And interestingly, there are times when appalling reviews actually generate substantial box office returns as audiences go to see if it 'really is that bad'.

Distribution

Once a film has been made, the distribution company (which may, of course, be part of the same vertically or horizontally integrated organisation) starts to promote it. In some instances, this marketing will start before the film is released, to generate 'pre-audiences' and heighten audience expectations – a long-awaited sequel is often marketed in this way. The film will be distributed first to cinemas and then released to video production, television companies, etc. There is usually a substantial length of time between the original cinematic release of a film and the video release date or the first television screening date. In this way, the distribution company can ensure maximum cinema audiences before the film is available in different media. Of course these timescales do change – and we all know of films which have gone 'straight to video' because they were unlikely to draw large cinema audiences!

Distributors

Major distributors include Buena Vista International, Columbia Tri-Star Films, 20th Century Fox Film Co, UIP and Warner Bros. As you can see, these are also the main film production organisations at present – the synergy of production and distribution being within the same organisation brings cost reductions but also ensures more effective control of the production and distribution process.

Exhibition

Exhibitors

Exhibitors are the places where the films are shown. Most of the blockbusters tend to be shown in multiplexes these days, with smaller cinemas not being able to show the films as quickly. The multiplexes will aim to show a range of new films across the different screens and will continue to show major blockbusters for a long period of time, gradually moving them to smaller cinemas within the multiplex as audiences diminish.

Weblink

It is very easy to find out about distribution and exhibition on www.imdb.com and you can often get useful box office figures etc as well.

Checkpoint 1

What do we mean by 'synergy'?

Many of the major multiplex chains are also part of the larger film groups, such as Time Warner, who own the Warner Village cinema chain. Again, this synergy allows the organisation to control exhibition as well.

Exhibitors want the rights for first screenings of blockbuster films and, ideally, to be able to restrict other cinemas from showing the film. UCI, for example, is part of the National Amusements organisation, which also owns Paramount and the Blockbuster video chain, so the integration of distribution, first exhibition and video exhibition is evident.

Video/DVD release

Once a film has been released in the cinemas, it will also, at a later date, be available on video and DVD. In the case of a successful blockbuster film, the distribution company will market this release date as strongly as the cinematic opening, knowing that they will generate a second audience in this way. Again, as with premieres, there may be free gifts, promotional articles or other artefacts available with the video and DVD to encourage audiences. While the video copy of the film will contain the film and maybe news of future releases (often from the same distribution company!) the DVD release is a much stronger marketing tool.

The first-generation DVDs of this kind were largely supplemented with footage which had not been used in the film but was still worth watching and a couple of brief interviews with cast and crew. Yet now, distributors are starting to realise the value of a well-made DVD with extensive supplementary material – especially to attract an audience which may well have seen the film at the cinema when it was first released. By adding a second DVD to the package (which is increasingly common), they are able to offer a lot of additional material to attract this increasingly demanding and sophisticated audience. Features which may appear on the second disk include:

→ trailers for the movie (often several trailers from different contexts)
→ behind-the-scenes shorts about how the movie was made
→ sequences which were edited out of the final version
→ interviews with cast and crew
→ simple games based on the film
→ supplementary materials such as screensavers or PC wallpaper
→ outtakes.

Checkpoint 2

How long does it usually take for a film to be released to video and DVD?

"Newscorp – they've got a studio. They've got a network, the Fox Network. They've got two cable outlets. So you know, they've got all these different arms to sell this stuff. Then they've got Sky Channel and they've got a satellite overseas and they own TV Guide. I mean, vertical integration is now – it's like Doctor Evil from the Austin Powers movies. You know, it's these tentacles that encircle the world.
When you've got Time Warner/AOL, Entertainment Weekly and CNN and Time Magazine and People and In Style, they get to flog movies in five or six different ways before the picture even comes out. And they're often pictures made by companies that they're in business with."

ELVIS MITCHELL, film critic, *New York Times* and *NPR*

Exam question   answer: page 119

Identify a contemporary film which has now been released on video and DVD. How was the film originally distributed and exhibited? (You may have to do some research before you can answer this part of the question – start by identifying who distributed the film and then study the DVD release.) How does this target the same audience, yet attract them to buy a film they have already seen?   (1 hour)

Pluralism

Just as uses and gratifications theory contrasts very neatly with effects theory in its analysis of audience, so Marxist theory and pluralist theory can be seen to be opposites. Whereas Marxist perspectives on the media centre on the belief that the mass media are a tool used by ruling bodies to maintain hegemonic control over the masses and a class-divided society, pluralists take a more flexible and open view, arguing for a classless society and media organisations which are responsive to an audience and economically determined. It is useful for you to understand both perspectives so that you are in a good position to use either theoretical base to support your arguments about a text.

A basic definition of pluralism

Pluralists view society as a system of competing groups and interests, none of them predominant. Media organisations are seen as enjoying an important degree of autonomy from the state, and control of the media is said to be in the hands of an elite who allow a considerable degree of flexibility in production choices. Audiences are perceived as capable of manipulating the media and having access to what Halloran calls 'the plural values of society', enabling them to 'conform, accommodate or reject'.

→ The pluralist tradition starts from the opposite belief to the Marxists and rejects the concept of a mass culture.
→ As we become more media-literate over time we become more demanding and more selective consumers, able to choose from a wide variety of cultural options.
→ Class distinctions are less important now in influencing the choices made by individuals.
→ High culture (ballet, opera, etc) now reaches wide audiences, and audiences which might have chosen high culture texts will now also choose soap operas, for example.
→ The individual has the freedom to make choices about what texts to consume based on their personal preference.
→ The media are seen as acting in the public interest as the **fourth estate**, keeping a close eye on abuses of power by politicians, large multinational corporations and others.
→ Pluralism allows for many different viewpoints to be explored in media texts – which may even be in conflict with each other (like party political broadcasts). There is no one hegemonic message to be conveyed through all media texts.

The role of the audience

If the audience does not like a media text they will simply not watch it or buy it. So ratings or circulation fall and the media organisations must respond by changing their output. The audience is vital to media

Checkpoint 1

What do we mean by audiences being able to 'conform, accommodate or reject'?

Links

This view of society relates closely to concepts such as uses and gratifications, so you may also want to look at pp. 90–91.

producers, who must therefore spend considerable amounts of time and research making sure their products will appeal to the audience. The audience is seen as **active**, selecting texts from a vast range, consuming them, not receiving them passively.

The media as fourth estate

There are various ways in which the first three estates of a democratic society are defined but essentially they are seen as the government, legal system and the church in various combinations. Each has power in our society (the church may be seen to have less power in Britain but media theorists here distinguish between the House of Commons and the House of Lords as separate estates and since the senior bishops are part of the House of Lords, the connection is valid). Pluralists argue that it is good that each of these systems is held in check by the other two because peace and democracy are thereby maintained.

The role of the media is seen as being that of the fourth estate, whose role is to keep the public informed about what is happening in each of the other three. The media fulfil this role by reporting political events in the newspapers or on broadcast television – by providing documentary programmes to report on key events, among other ways. For instance, BBC *Question Time* is seen as an important programme for giving feedback to the government about current issues and debates. Where there has been evidence of possible misconduct by the police or judges, for example, the media argue that it is their role to investigate and keep the public fully informed.

Of course, there are many cases when this is complex and the role of the media as fourth estate has been significantly affected post-Hutton by the debates which ranged around the Iraq war and the roles of the government and the BBC.

Limitations of pluralism

One of the main problems with pluralism is that it starts from two assumptions – it assumes an active audience which dictates the texts on offer because media organisations respond to audience demands, but it also assumes an economic basis for production. At times these two assumptions can be seen to be in conflict. After all, it is not always possible to create the texts which the audience wants and make profitable programmes, for example. A magazine may be demanded by an audience but unless that audience is suffiently large to mean that sales are economically viable, the magazine may not be produced. Of course, there are subsidies and grants available to make these texts – e.g. the licence fee supporting the PSB charter of the BBC.

Checkpoint 2

What is the fourth estate?

Example

The Hutton report (prompted by the dispute between the BBC and the government after the suicide of Dr David Kelly) was a key moment in the relationship between the media and the government because the BBC were heavily criticised for acting too strongly and abusing their role in society by deliberately trying to put the blame on the government.

Check the net

You may like to research this case using the web archive at Guardian media (http://media.guardian.co.uk/)

Exam question   answer: page 119

Look at the TV schedules for this week. How might a pluralist interpret the range of programmes on offer? What evidence is there of the media acting as the fourth estate during this week?   (1 hour)

Answers
Media issues

Effects theory

Checkpoints

1 The culture industry is a way of describing media production in economic terms, emphasising that culture is a commodity (or thing) which can be bought.

2 'Desensitised' means that we have become so exposed to something that it no longer has an effect on us. So, because we are exposed to greater levels of violence in the media we have become desensitised and accept this as the norm.

Exam question

This is quite a wide-ranging question. Logically, the first place to start is to identify what hegemony might be being promoted (e.g. cultural expectations about the police force or attitudes towards asylum seekers) and then to relate these to the actual scheduling.

Stronger answers will move forward to consider the role of the different TV channels in promoting hegemonic values and should also move beyond the simplistic PSB=hegemony and commercial=profit-led split to analyse the positions a little more carefully. Answers taking in a lot of wider contexts post-Hutton may well argue that the BBC does not seek to maintain the hegemony but to subvert it; others could evaluate the role of Murdoch in Sky channels and how he influences them in relation to hegemonic order.

Reception theory

Checkpoints

1 The context of consumption is where you are and who you are with when you consume a media text – if you are watching a romantic film with your parents, for example, you are likely to respond differently and say different things than if you were only watching with your boy/girlfriend!

2 'Dominant reading' means preferred reading – the one intended by the media producers. 'Negotiated meaning' means that the audience will accept some parts of the preferred reading but will modify it to suit their own purposes. An 'oppositional reading' means that the audience will reject the preferred reading and receive the text in a very different and usually negative way.

Exam question

Obviously, the structure of the response here will depend on the programmes chosen but it should be fairly clear how these readings can be differentiated and how different audience groupings might respond to the texts. Stronger answers here would move beyond simply identifying these different readings to also explore the reasons behind them and how the readings might be affected by the context of consumption, for example. They should be able to consider the nature of the active audience and may even be able to relate this to the uses and gratifications for these audience groups for these two texts.

The two texts chosen could be contrasting or similar (e.g. two programmes starring the same person) or two lifestyle programmes on different channels, for example. There may not be any links between the audience groups, of course.

Uses and gratifications

Checkpoints

1 Bulmer and Katz suggest that we choose a media text for diversion, personal relationship, personal identity or surveillance reasons.

2 Cultural codes are codes and conventions which relate to particular cultural groups in society. So particular cultural groups will receive a media text in line with these cultural codes.

Exam question

This is the counterargument to the argument presented for the effects theory section. It is worth realising that you can analyse the same material from both oppositional views because that demonstrates a good understanding of these two contrasting schools of audience theory.

Here, you would be looking to see how audiences might construct a schedule for their evening viewing in line with their own particular uses and gratifications, for example, choosing programmes possibly with very differing ideologies as part of a more 'pick and mix' approach. This assumption that they can choose their programming in this way also assumes that they can evaluate and reject ideologies which they do not support – thus rejecting the influence of hegemonic values you wrote about for the effects theory section.

Again, as before, stronger answers will move beyond basic examples of this kind to consider how the schedules themselves persuade viewers to 'surf' in this way (or not, as the case may be) and therefore also the different nature of programming across channels and slots – further suggesting a good deal of audience choice.

Narrative theory

Checkpoints

1 Propp argued that conventional narratives are constructed like fairy tales.

2 Lévi-Strauss analysed narrative in terms of binary oppositions (good/bad, young/old).

3 Barthes identified particular codes at work in narrative texts – the best known is the 'enigma code', which intrigues an audience so they want to find out more about events.

4 Todorov argued that narratives are constructed in sequence – an equilibrium is established at the beginning and then disturbed. The process of the text is to restore the equilibrium or form a new equilibrium.

Exam question

Deconstructing a trailer in terms of these four basic narrative theories should be fairly straightforward and you should find

plenty to say. In some ways, some of these sample essays are very straightforward whereas real exam essays can be more complex because you need to write across a lot of different theoretical positions at once.

Here, the strongest answers will allow time to explore and justify which of these four narrative theorists is most useful when analysing this trailer and remember to justify why this is the case.

Representation and regulation

Checkpoints

1 A stereotype is a standardised (often negative) representation of a type or group.
2 A self-regulatory regulatory body is set up by a media sector from members of that sector and is usually voluntary. OFCOM is neither self-regulatory nor voluntary.

Exam question

This answer needs to be contextualised first in terms of which sitcom and soap opera you have chosen. Some comments about why you have chosen these particular texts in relation to this question would probably be very helpful.

You should then pick at least two characters from each programme and explore the degree to which they can be defined as stereotypical or exhibit stereotypical characteristics. You don't need to explain what a stereotype is or list the characteristics of a stereotype without relating these to the specific texts with detailed examples. Examiners won't give you credit for having learnt about stereotypes, only for applied understanding showing that you can use theoretical concepts to structure your detailed analysis of the texts.

News values

Checkpoints

1 You should be familiar with these news values but there is no set hierarchy for them.
2 Gatekeeping is the process by which editors control what news stories are actually released to the audience and how/when they are released.
3 Bias means favouritism – for example, most newspapers are very biased because they have a particular political position.

Exam question

It would be a good idea to complete a table as part of your research for this essay, listing each of the items in the bulletin. (By all means separate off the headlines section, main section and concluding section to consider each of them separately.) For each item record how long it was given, what supporting material was used (e.g. pictures), what sort of language was used (e.g. workers demanding and managers offering) and which news values seem to justify the inclusion of that story. Once you have completed this analysis you should also be able to identify what kind of

gatekeeping has gone on and also what bias is apparent (or implicit) in the broadcast.

Stronger answers, as always, will integrate theory and evidence, making good use of the theory to support the points they make in relation to the institutional position of the broadcaster and the wider context of contemporary society.

Audience profiling

Checkpoints

1 Demographic audience profiling groups audiences by age, gender or location, for example – as groups of people.
2 Psychographic profiling groups audiences by likes and dislikes, e.g. aspirers or carers.
3 Advertisers like to create these niche markets because, as people identify with the niche, they aspire to buy the products associated with the lifestyle.

Exam question

Probably the easiest way to approach this type of question is by working backwards. If you start by simply identifiying the target audience from some of the groupings you have been given and then go on to analyse the ideologies constructed in relation to this audience, you will probably end up repeating yourself. It's far better to start by identifiying the ideologies constructed within the campaign and then relate these to the values and attitudes of specific audience groups – in this way you can talk in far more detail about the methodologies used by advertisers to construct audiences and reach them, which is the most interesting aspect of the question.

Advertising

Checkpoints

1 Intertextuality is where a text makes use of a different media text to create meaning in the text. So, for example, an advert may make intertextual reference to a well-known film in order to create meaning for an audience.
2 Product placement means that the makers of a media text will be paid to use the product in the text. So, for example, Aston Martin used to pay a lot of money to ensure that James Bond drove an Aston Martin.

Exam question

This is another fairly straightforward question, which is basically asking you to pick a campaign to study in detail and move that study beyond simply deconstructing the media language used in relation to the target audience. Here you need to explore the ideologies constructed and how they almost create a lifestyle grouping; how the campaign reflects the institutional context (which can be complex if the advert has been contracted to one of the major advertising companies, for example) and finally to consider how the campaign succeeds in generating desire in the target audience. This may lead on to discussion about the active audience and to what extent people generate these lifestyle labels or adopt them once they have been defined by

advertisers. You might also discuss the role and function of advertising in general and the amount of power it has over consumers.

Postmodernism

Checkpoints

1 A metanarrative is one which collects together a range of smaller narratives in order to make meaning. For Lyotard, therefore, any religion or political belief system such as Marxism is a metanarrative.
2 Hyperreality is a representation of a reality which has almost become more real that the reality it was originally intended to represent – e.g. news reports.

Exam question

This question relates directly to Baudrillard's assertions about the 'reality' of news reporting against actualité. A good starting point would be to explore this relationship on a basic level – for example, by directly comparing how one story can be edited very differently for different contexts (ideological, institutional or conceptual) and then to move on to consider the relationship of the simulacra with the actualité. This is a complex area so you should make detailed notes before you start and remember that you must link all your arguments back to specific examples. You may find it helpful to use two particular texts here, e.g. the evening news from BBC One and, say, CNN. (Note that even the concept of 'evening news' becomes postmodern in this context because of the time differences between London and Washington.)

Genre theory

Checkpoints

1 A media-saturated society is one which is absolutely full of media texts of one kind of another. Our society is so media-saturated it is almost impossible to avoid consuming a range of texts every day.
2 The semantic elements of a text are the media language, representations and codes and conventions. The syntactic elements are those which structure the text, such as the narrative or ideologies.

Exam question

This has been included here as a typical example of a general question, especially at A2 level. It would be easy to answer a question like this in a very general way, referring to a range of texts and genres and making a lot of sweeping statements about the relationship between audience and genre.

In fact, to produce a strong response to this question you need to apply it to detailed study of a specific genre. The key is to consider the relationship between audience and genre. Very strong answers would be able to address this question across two genres – possibly even being able to argue that for one of the genres (such as science fiction films) the generic elements are fundamental, as the audience is more interested in the genre than the particular text (a very good example here is the Bond formula).

Equally, for other genres, the opposite may be true. Comedy is a good example here because comedies need to be fresh to be successful. Of course, if you follow comedy back, it is clearly formulaic (think of Laurel and Hardy) yet contemporary comedy as a genre needs to move beyond generic expectations in order to win audiences.

Star theory and auteurs

Checkpoints

1 A media mogul is a media owner who owns a very large media organisation (or group of organisations) and has significant influence in the media, often well beyond the bounds of their own organisation.
2 An auteur is a director who so dominates the film making process that it is appropriate to call them the author of the film. The auteur theory holds that the director is the primary person responsible for the creation of a motion picture in his or her own distinctive, recognisable style.

Exam question

This is an ideal question to answer on your favourite film director, e.g. Welles or Tarantino. It is not a question which can be answered successfully without a lot of research and preparation but is very straightforward once that has been done, as long as you remember to support all your points with detailed evidence from the text.

By identifying key signifiers of the auteur and exploring how they create meaning in a particular way, constructing a particular ideology and probably making use of particular media language (which may or may not be generic to this director) they are able to make meaning in a significant way. Often the first way to identify a film made by an auteur is that you are aware of the director of the film almost as soon as you have heard of the film and the ideology of the director frames your likely reception of the film.

Gender studies

Checkpoints

1 The male gaze is Laura Mulvey's explanation for the conventional structure of Hollywood narratives. She argues that action/adventure films in particular are constructed in such a way as to appeal primarily to a male audience. So, the hero must always be someone that this audience can aspire to be and the female characters must be very attractive (eye candy) but play passive or subservient roles to the action hero.
2 'Eye candy' refers to a character or presenter in a visual media text who is there primarily because they are attractive. They may have no significant narrative function or ideology – they are just there to be visually appealing.

Exam question

Again, your response here will differ depending on what the five top films are when you do this. If you are lucky enough to have a week where a non-Hollywood movie is doing well,

you should be able to make some useful contrasts about the use of the male gaze to sustain the audience. If you are looking at a week where conventional Hollywood texts predominate, your answer will probably be structured rather differently as you may well write about the hegemonic control of the male gaze and the influence this has, not only on the construction of these films, but also the ideologies being conveyed to the audiences and, indeed, the audience expectations. Stronger answers will be able to move beyond this perhaps to consider the influence of Hollywood on young women, and also in relation to feminist theory about gender construction, for example.

Marxism

Checkpoints

1 The proletariat are the workers (or the masses) and the bourgeoisie are those who employ them. The petty bourgeoisie are those who employ others but also work themselves.
2 For Althusser, the ISAs and RSAs are a means of maintaining bourgeois ideologies in society. The ISAs do this by indoctrinating the masses through religion, education, media, etc and the RSAs force us to comply.
3 Montage editing as employed by Eisenstein is a way of putting together a sequence of shots in a film to create meaning through the interconnection of the shots and the way they establish ideology and narrative. It does not rely on realist construction.

Exam question

You need to be sure to write in enough depth in response to a question of this type. In a limited time frame you can't possibly write in detail about the whole of a film like this, so you need to have detailed notes about specific sequences in the film such as the Odessa steps sequence. Once you have identified particular sequences for study, this is a fairly straightforward analysis question, relating media language (editing) to ideology and function, which you can do with plenty of textual examples.

Distribution and exhibition

Checkpoints

1 Synergy means the relationship between two different media organisations, which helps both to reach wider audiences.
2 Usually this timescale is six months to video release, one year to satellite film channels and two years to terrestrial television channels.

Exam question

This is another example of a question which can lead to very general answers – despite asking you only to write about one film. The danger here is that you simply identify what extra features have been added to the DVD version of the film almost as a list and summarise your thoughts about why this is done as 'to attract the audience'. In fact there are a lot of debates to be raised here about the ways in which DVDs are now marketed – is this essential to prompt audiences to watch the text a second time? After all, video copies do not also offer these special features!

Stronger answers might well look further at the whole way that films are marketed once they are released for purchase or rental and also consider the complex job for the institutions to sell the same thing twice. This would lead to some interesting comments, perhaps about vertically and horizontally integrated companies and also synergy in terms of promotion.

You might also want to reflect on how much use is actually made of these extra features, for example. What about films which are re-released many years after original cinematic release? Is there merit in the bonus features here as well, or not so much? Why are distributers prepared to pay so much extra to put these bonus features onto the DVDs?

Pluralism

Checkpoints

1 Audiences being able to conform, accommodate or reject a text is simply a different way of presenting David Morley's explanation about audiences accepting dominant, negotiated or oppositional readings of a media text.
2 The fourth estate means the way the media act almost as public watchdogs, making sure that government behaves properly, for example. It suggests that the media may be the tool by which the people can influence the government.

Exam question

There are two aspects to this question. The first part asks you to explore the range of texts available and how the audiences for these vary extensively across the schedules. The second part of the question asks you to look at the range of programmes, such as *Question Time*, where the people, via the media, have a chance to influence government. Why do you think the overall balance of programming is this way?

Media issues and theory
Revision checklist

1	Understand effects theory, the hypodermic needle model and cultivation theory	Confident	Not confident. **Revise** pages 86–87
2	Understand uses and gratifications theory	Confident	Not confident. **Revise** pages 90–91
3	Understand reception theory and the importance of the work of David Morley	Confident	Not confident. **Revise** pages 88–89
4	Understand how audiences are profiled by media organisations	Confident	Not confident. **Revise** pages 98–99
5	Understand bricolage, hyperreality, simulacra and metanarrative in relation to contemporary texts	Confident	Not confident. **Revise** pages 102–103
6	Understand the basic principles of Marxist theories in relation to the media, especially hegemony and debates around violence in the media	Confident	Not confident. **Revise** pages 110–111
7	Understand how genre is used to define and control media texts	Confident	Not confident. **Revise** pages 104–105
8	Understand the role of the male gaze in media texts and the ways media texts are constructed in a patriarchial society	Confident	Not confident. **Revise** pages 108–109
9	Research individual stars and auteurs in relation to particular texts and their industry sectors	Confident	Not confident. **Revise** pages 106–107
10	Identify stereotypes used in media texts and how representations in media texts are regulated	Confident	Not confident. **Revise** pages 94–95
11	Understand how news values affect the selection and construction of the news	Confident	Not confident. **Revise** pages 96–97
12	Understand the four main theories of narrative structure	Confident	Not confident. **Revise** pages 92–93
13	Understand the importance of distribution and exhibition to the film industry	Confident	Not confident. **Revise** pages 112–113
14	Understand pluralist theories in relation to the media	Confident	Not confident. **Revise** pages 114–115
15	Identify different forms of advertising such as overt and covert advertising, sponsorship, product placement and viral advertising	Confident	Not confident. **Revise** pages 100–101

Here we look at some of the most important debates that you should be familiar with when you are studying Media Studies. You should also make sure that you are familiar with the jargon used in these debates, remembering that use of media terminology is expected at A2 level in particular. You will not be expected to cover all the research areas so you need to check with your teacher which ones you are studying.

Of course, we have not covered all the possible topics, but from these examples you should be able to see how to approach the research area and what you might want to research. You will need to do your own thinking about these areas so these pages are only starting points. Your research will need to focus on specific texts; you need to discuss your choice of texts carefully with your teacher as well.

Exam themes

→ Media regulation

→ Types of media text

→ Changing ownership, consumption and technology in the information society

Topic checklist

○ AS ● A2

	OCR	AQA	WJEC
Documentary forms	○ ●	○ ●	○ ●
Censorship	●	●	●
Media ownership	○ ●	○ ●	○ ●
Globalisation	○ ●	●	●
New media and the information society	○ ●	●	○ ●
Cyberspace and identity	○ ●	○ ●	●
Cult of the celebrity	●	○ ●	●
Computer games	○ ●	○ ●	●
Popular music	○ ●	○ ●	○ ●
Crime and violence	●	○ ●	○ ●
Politics and propaganda	●	○ ●	○ ●
Sport and the media	●	●	○ ●
Women and film	●	●	●

Documentary forms

Checkpoint 1

What is actuality filming?

Check the net

There is a copy of the essay Flaherty wrote about making *Nanook of the North*, which you can read at http://www.cinemaweb.com/silentfilm/bookshelf/23_rf1_2.htm

Check the net

Nick Broomfield has his own website for his work, which is at www.nickbroomfield.com

"[Broomfield] works by zooming in on the inessentials as the means of getting to the heart of the matter. He leaves in what conventional documentary makers would edit out."

John Carlin, *Daily Telegraph*

There are many different types of documentary and many documentary makers. A documentary is usually defined as being a non-fictional representation of reality, which has a clear narrative structure and makes use of a range of devices to communicate meaning to an audience. A documentary will usually focus on a single issue or story and may present alternative views or a single vision. A conventional documentary is intended to inform and educate as well as entertain. Hybrid documentaries make use of the documentary form mixed with some other genre elements – usually for entertainment more than information or education.

Early documentary

John Grierson created the term 'documentary' to refer to 'the creative treatment of actuality' and cited Robert Flaherty's ***Nanook of the North*** as the model: a film that includes re-enacted sequences and a somewhat romanticised 'noble savage' view of its protagonist. Narrative and character were used to establish reality in a way which was very different from the 'actuality filming' which had comprised documentary to that point. Flaherty is generally regarded to have established the rules of early documentary, Grierson to have formalised the genre.

In 1934, John Grierson became head of the GPO Film Unit and made some of the earliest documentaries in this country. Film, according to Grierson, had to be entertaining as well as informative. It could directly engage audiences but should aim to move beyond actuality to represent the real behind the actuality. He argued that actuality had to be shaped and treated in order to create meaning for an audience.

One of his most famous documentaries was ***Night Mail***, famously linking the poem 'Nightmail', recited by W H Auden, with footage of the night mail train travelling between London and Edinburgh. He sought to use the art of his documentary to celebrate the men who made the Royal Mail train run so well. However, critics have commented that the train itself is celebrated so much that the workers become secondary.

Direct cinema

Direct cinema is a much more conventional form of documentary and seeks to 'record' events – mediating those events as little as possible when constructing the documentary. Fred Wiseman, Michael Moore and Nick Broomfield are all associated with this genre of documentary.

Direct cinema seeks to make films in a direct, immediate and authentic manner. This genre constructs a narrative, even though the impression is given that the phenomena are recorded exactly as they happen. This has proved a useful device in the delivery of contemporary news programmes, though these examples are much more canned than direct.

The film maker is invisible and seemingly uninvolved (although in full control of the editing process, of course).

Cinema verité

The style of cinema verité originated in France during the 1950s and 1960s. Verité is **minimalist** and conveys the sense that the viewer is

given a direct view of what was actually happening in front of camera on the day of filming. Absence of TV lighting and the rarity of interviews emphasise this 'reality'.

Verité uses small hand-held cameras, and actual homes and surroundings as the location for a film. One convention is to tape-record actual conversations, interviews and statements of opinion made by real people and then find pictures to illustrate the sound recordings. The final production is put together in the editing room (which is also true of fiction/fantasy films). The film maker's goal is to show life as it really is, using the film as his artistic medium.

In fact, the input of the film maker is often at the root of criticisms of verité. Documentary makers inevitably manipulate reality to such an extent that the film can be very different from the reality under examination. (Think of the amount of editing which goes on when a documentary maker has been filming a subject for a year but must produce a programme which is only an hour long.)

Cinema verité is part of the broader artistic tradition of **realism** and the cinematic tradition of documentary film making. The film maker will usually have a social conscience and sometimes a political agenda. His purpose is to enlighten his audience, to show them the truth as he sees it, so they will have the information they need to live better lives or, in some cases, to take political action to right the wrongs the film maker often exposes.

The jargon

Minimalist means as sparse and simplistic as possible without losing meaning.

Lifestyle programmes ●●●

Lifestyle programmes present information or explore topics in a superficial way – depth of insight and critical analysis/commentary are rudimentary or absent. They are usually **aspirational** and focus on accomplishing practical objectives and thus, on viewers' aspirations. Lifestyle programming often contains elements of other genres (e.g. travelogue, how-to, reality television) which means lifestyle programmes are usually referred to as **hybrid** documentaries. It is a form intended as **entertainment** – information is secondary and education superficial. A similar genre of hybrid documentary is human interest programmes which act as a 'window' for an audience to view other places, people or events – still as entertainment and in a very positive way.

Checkpoint 2

Why are lifestyle programmes generally seen as aspirational?

Mocumentary ●●●

Mocumentaries are films and television programmes which look and sound like documentaries, but are not factual. They use the same codes and conventions as documentary, such as an authoritative narrator, 'real' footage of events, archival photographs, interviews with apparent experts and eyewitnesses and so on. Because they demonstrate how easily all of these codes and conventions can be faked, mocumentaries can often cause us as viewers to consider why we place so much faith in documentary itself.

Exam question   answer: page 148

Watch a documentary by Nick Bloomfield or Michael Moore. In what ways do they construct the documentary to promote an ideology?   (1 hour)

Censorship

Checkpoint 1

Is the PCC a statutory regulatory body?

It is important to be aware of legislation affecting media organisations in Britain. Although the legal situation is different for other countries, it is obviously the UK which is of most interest to you. The most important bodies to know about are the PCC and the BBFC. There are also several laws which impact on media texts and these are also mentioned here. You are unlikely to meet material which breaks these laws but you should be aware that they exist.

The Press Complaints Commission (PCC)

The PCC was founded in the early 1990s to create a code of practice for journalists and photographers; a press complaints tribunal system and legislation relating to intrusion of privacy. The PCC is self-regulatory. In other words, it is a voluntary body, run by the media organisations, and media organisations must choose to sign up to the PCC code. The code covers issues such as:

→ accuracy
→ the right to reply (for people written about)
→ not misleading readers or misrepresenting people
→ privacy and harassment
→ discouraging cheque book journalism
→ being sympathetic to victims and also to the family and friends of people in the news (e.g. those on trial) – including retaining their right to privacy
→ appropriate and discreet reporting of stories with children involved or involving victims of sexual assult, for example
→ acting 'in the public interest'.

Action point

Do you think that it should be compulsory for news organisations to belong to the PCC?

The important factor about the PCC is that it is voluntary. There is no censorship of news and no 'government' newspaper in this country (whereas in many countries there is only the state news service). There are, however, laws which affect newspapers, broadcasters and all media organisations, some of which are listed here:

→ **The Official Secrets Act**
  This is used to ensure that anyone who works for the military, government or police may not speak to the press about any aspect of their work without permission. The OSA can also be invoked to allow the government to serve a **D notice** on news organisations to prevent them reporting about a war, for example.
→ **Libel law**
  If a media organisation prints or broadcasts a story about someone which damages their reputation, character or lifestyle (e.g. if they lose their job because of something in the papers), the person has the right to sue the organisation for libel. This will usually involve the claiming of a substantial amount of money to make up for the damage caused.
→ **The Cinematograph Films (Animals) Act 1937**
  It is illegal to show any scene 'organised or directed' to involve actual cruelty to animals.

Don't forget

You do not need detailed knowledge of media legislation for your essays. It is important to be aware of what legislation there is but you will not be expected to explore this in any detail unless you are researching censorship as an individual study.

→ **The Protection of Children Act 1978**
It is illegal to show indecent photographs of a child.

→ **The Obscene Publications Act 1959**
It is illegal to show a work which is obscene. A work may be found obscene if it might disturb and corrupt a significant proportion of those likely to see it.

→ **Race Relations Act (1976)**
It is illegal to publish or disseminate materials, such as leaflets and newspapers, that could be considered racist in any way or which may incite racism.

→ **Sex Discrimination Act (1975)**
Men and women must be represented equally and fairly in the media.

The British Board of Film Classification ●●●

The BBFC is an independent, non-governmental body, which has exercised responsibilities over cinema since 1913, and over video since 1985. Moving image material is submitted to the BBFC for rating and the BBFC will award a ratings certificate which identifies what audience age range the material will be suitable for. These categories are:

 Films rated Uc are primarily for pre-school children.

 Films rated 'Universal' are suitable for all ages.

 Parental Guidance films are deemed suitable for children but there may be scenes parents would not wish children under 12 to see.

 No-one younger than 12 may see a '12A' film in a cinema unless accompanied by an adult. No-one younger than 12 may rent or buy a '12'-rated video.

 No-one younger than 15 may see a '15' film in a cinema. No-one younger than 15 may rent or buy a '15'-rated video.

No-one younger than 18 may see an '18' film in a cinema. No-one younger than 18 may rent or buy an '18'-rated video.

'R18' videos may be supplied only in licensed sex shops to adults of not less than 18 years. The 'R18' category is a special and legally restricted classification primarily for explicit videos of consenting sex between adults. Such videos may be supplied to adults only in licensed sex shops. 'R18' videos may not be supplied by mail order.

> **Checkpoint 2**
>
> Are you allowed to watch a '15' rated film at home if you are under 15?

> **Action point**
>
> Do you think it is a good idea that we continue to classify broadcast texts in this way?

> **Exam question**   answer: page 148
>
> Will we ever reach a point where, as an audience, we are so desensitised that we don't need the BBFC any more?   (1 hour)

Media ownership

Media ownership is a key issue for Media Studies. You need to be aware of the differing types of media organisation and how they are structured. Although you will probably undertake research into specific media organisations in class, we will not be using specific case studies here because media ownership changes so quickly. Make sure that you check information about ownership of media organisations you are researching very carefully, for this reason.

Checkpoint 1

What is a conglomerate?

Types of media organisation

→ **Monopolies** – a monopoly means that a single media organisation dominates production and distribution in a particular industry, either locally or nationally.
→ **Oligopolies** – an oligopoly is when a small number of media organisations dominate an industry, either locally or nationally.
→ **Conglomerates** – a conglomerate is a collection of companies owned by a single institution. These need not all be within the same industry. This diversification allows protection against a single part of the conglomerate failing.
→ **Multinationals** – a multinational organisation is one with institutions in more than one country. This can have economic advantages as well as giving better audience reach and diversification.

Control and domination

Larger media organisations are able to control the making of the product, distribution, marketing and often even the outlets it is sold in. For example, Warner Bros can make, distribute, promote and exhibit a blockbuster without needing any outside organisations. They can even control video and DVD releases of the film.

This can have big advantages for the organisation, ensuring significant profit levels, but it can be less advantageous for the audience. In this marketplace there is less space for alternative or independent films, for example, or ones which challenge the cultural hegemonies maintained by Warner Bros. The media we receive can be too closely controlled by a single organisation. Key institutions in this country at the moment which you may well research as part of your study are:

→ Sony
→ Vivendi Universal
→ Bertelsmann
→ News Corp
→ Viacom
→ Walt Disney
→ AOL Time Warner.

Monopolies Commission

In the UK, the Monopolies Commission exists to ensure that a single organisation in any industry sector cannot simply buy out all the

Check the net

A good starting point to find out more about each of these key conglomerates is to look on their websites for information.

competition and gain a monopoly in that area without very good reasons. It has looked into a number of media organisations over recent years to try and prevent a media monopoly forming in any one media or communications sector.

Vertical and horizontal integration

Action Point

A **media mogul** is usually in charge of a significant media organisation. Which media moguls can you think of in the UK?

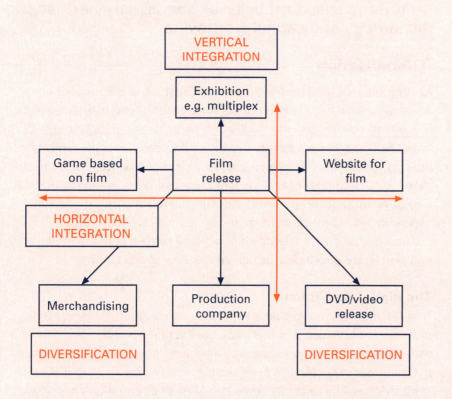

So, in the case of Warner Bros say, **vertical integration** means that they can control all stages of production, distribution and exhibition for a new film release. **Horizontal integration** means that they also own the institutions to produce the game based on the film and the promotional website, for example. **Diversification** into other industries means that they can also produce and market the DVD/video release and merchandising related to the film, e.g. T-shirts. **Convergence** refers to the way increasing numbers of smaller media organisations are becoming part of these larger organisations.

Checkpoint 2

What do we mean by diversification?

Checkpoint 3

What do we mean by convergence?

Synergy

Where institutions are not owned by the same organisation but work together for mutually beneficial purposes, this is called **synergy**. In a media context, this can relate to complementary organisations working together – for instance if Warner Bros were to license a different institution to produce the game based on a film because they did not have the in-house resources to do it themselves.

Exam question   answer: page 148

Choosing one of the seven multinational media organisations above, write about the ownership profile of the organisation. You should research the organisation carefully before beginning your essay.   (1 hour for essay)

Globalisation

Globalisation is becoming an increasingly important concept for media students. Continuing technological advances and growing global conglomerates are changing the face of the media. You should be aware of the ways these changes are happening and some of the issues which they raise. Globalisation centres on the increased mobility of goods, services, labour, technology and capital worldwide.

Globalisation

As international trade borders come down (think of the number of countries which have joined the EU, for example), media organisations are able to reach increasing audiences. Large media organisations have the resources to expand globally to become global players, often pushing home-grown media organisations out of business and replacing them. AOL Time Warner and Vivendi Universal are now among the largest institutions in the world. Advanced telecommunications and expanding global media markets mean that national and international cultural traditions are merging into a global culture and new media and genres are developing as part of this new global culture.

The digital revolution

New technologies mean that we socialise differently (e.g. using text messages to organise evenings out), bank and shop online and expect a wide range of television channels. These new media are reshaping our lives and our expectations. For example, we are able to shop globally, send video emails to family abroad and few of us could manage now without a mobile phone.

Technological convergence

As the technology continues to evolve and media organisations continue to converge, we are moving towards an environment in which all our media – from television to the telephone – will be accessed via a single device, creating new forms of interaction across and among media. Each new generation of mobile phone or PDA, television or computer is able to access an increasing range of media. We are increasingly adept as consumers and citizens at using one medium in relation to another, choosing which technology is best to receive or transmit a particular form of information, and transforming media content to express our own ideas better.

Cultural convergence

With this increased access to global media and increased demand for consumer choice, the largest media organisations are supplying progressively more of the media we consume. Inevitably, much of this is US-centric and globalisation has been termed the 'McDonaldisation' of the globe. Our cultural perceptions are primarily constructed by the media – so we receive the values and ideologies they establish. This is sometimes called **cultural homogenisation**.

Action point

How much of the media you consume is global? Do you watch US TV programmes or visit US websites, for example? Do you consume media from other cultures as well or does the US dominate?

Checkpoint 1

What is cultural homogenisation?

Cultural imperialism

Critics of this cultural dominance refer to it as 'cultural imperialism', referring back to the times of the British Empire, when the Empire ruled a large proportion of the developing world and forced British values and ideologies upon it. Today's critics argue that the current US media domination amounts to cultural imperialism because it forces US culture on us through our media consumption. This is obviously damaging for national media organisations and especially for small independent organisations. The documentary film maker Michael Moore explores this in his films – and has even won an Academy Award for them!

Cultural imperialism through news globalisation

One current debate about the effects of globalisation revolves around the global news system that has developed as a result of technological advances. Some critics argue that national broadcasters are capable of and responsible for selecting and repackaging information to best suit their conceptions of domestic needs. Others argue that the US monopolises the global information system, dominating nations that wield less power in it. News is culturally dependent but is also product and producer of that culture. Is the dominance of a few global news organisations a good thing or dangerous?

The global village

Marshall McLuhan, a Canadian media critic, famously stated in the 1960s that the world had become a 'global village'. The media play a vital role in creating this. We can view events live as they happen. Whether it is bombs being dropped upon Baghdad or France being defeated in a World Cup, many people around the world can share the same moment.

Critics argue that the media reflect and create the social and cultural world we live in because the media producers construct our views of all these global events and therefore construct our values and ideologies. Reception theory contradicts this by arguing that the audience is active in this process and therefore the media producers only produce material which reflects the values and ideologies of the perceived audience.

Media is now an essential part of many people's everyday life. McLuhan stressed the role of the media in creating 'the mass' as opposed to the separate individuals of 'the public'. He also made very sweeping statements that electronic technology 'fosters and encourages unification'. Does the dominance of Western media, especially American products and styles, around the world create this kind of global shared meaning?

Checkpoint 2

What is cultural imperialism?

Example

Eric Schlosser, a US journalist, has written a very well-known book about this all-pervading US cultural imperialism, called *Fast Food Nation*, which you can borrow from your local library.

Television has become our eyes, the telephone our mouths and ears; our brains are the interchange for a nervous system that stretches across the whole world.

Benjamin Woolley

Links

Remember to think about news globalisation in relation to news values (p. 96).

Checkpoint 3

What is the 'global village'?

Exam question   answer: pages 148–149

Some critics argue that the growing Disney brand is a good example of cultural imperialism at work. Do you agree?   (1 hour)

New media and the information society

As new media change and shape our cultural perspectives, society is moving from an industrial to an information-based society. This is sometimes called the 'knowledge society' because information or knowledge is now seen as more important than production. Our world is media-saturated and our perceptions controlled by the media.

Checkpoint 1

Why is this the information age?

Check the net

Project Gutenberg is at
http://www.gutenberg.net

The information society

Western society has moved from the industrial age to the information age. Whereas once society was focused around large factories and Marx's view of the mass audience was formed by these factory contexts, the information age is different. The emphasis now is on finding and distributing information and media audiences are **niche** audiences (think of the growing magazine market). It is important to remember, however, that this has not led to an increase in smaller, independent media producers, but rather the increase of global media corporations, dominating the industry. Think, for example, of the large publishers which now control the majority of magazine publication in this society. They have access to the information and they are able to package the same information differently to reach different niche markets.

Information-rich with new media

It is the new media which have moved us to the information society. With the rise of the Internet, increased access to global televison and radio services, 'live' news reporting from around the globe and growing ranges of information sources (magazines, weblogs, ezines, promotional magazines . . .) we have access to far more information than ever before. Some critics argue that because so much information is now available to us, we are demanding still more information and that ultimately all information will become accessible. **Project Gutenburg**, for example, is an Internet site that has many books and other materials available for free download, such as the full works of Shakespeare, Tolstoy, Dante and many others. The database can be searched from anywhere on the Internet.

Digital divide

Another key issue related to the availability of these new media and information is that many people are excluded by economics, language and subculture from this information. There are several key dividing factors:

→ socio-economic status – you need money to have Internet access
→ developing countries often do not have the infrastructure for easy access
→ non-English speakers are often excluded because English is used throughout cyberspace
→ much content is US-centric – so US values and ideologies are communicated globally.

Marshall McLuhan

McLuhan charted the way that the media have shaped the way we receive and understand information.

→ In the **tribal** era information was received verbally and therefore information was narrative-based (story or poetry) and mythology and fable structured understanding of our world. Information was presented in an engaging way to keep an audience's attention.

→ In the **Gutenberg** age, the invention of printing encouraged people to think in straight lines and to arrange their perception of the world in forms convenient to the visual order of the printed page. Information became individual rather than shared (reading a book).

→ In the **electronic age of retribalised man** information is multisensory in that we receive it visually and aurally and we can select our information sources from a vast array. In addition we do not need to receive information as a linear narrative, expecting multilayered narratives and hyperlinked, interactive information sources. Once again we are expecting to be entertained as well as informed.

The medium is the message

This led to one of McLuhan's most famous statements that it is the medium – not the content – of the message which influences social and cultural progress. For example, the nature of the interactive webpage has shaped the way we expect to interact with screen-based materials – we expect lots of visual content, extensive hyperlinks, possibly animation and sound. In other words, our expectation of the information is shaped by our expectations of **how** it is delivered. In the same way, we demand particular genres of television because we appreciate the genre more than the particular programmes. Pleasure comes from watching lifestyle programmes, rather than selecting particular programmes to watch.

Information and entertainment

As the range of information available to us increases, our expectations of entertainment increase as well. The increase in narrowcast television channels such as the Documentary Channel or the History Channel reflects this – as more information becomes available, so we demand more information, presented with increasingly high production values and entertainment expectations.

Links

This links very closely with Dyer's theories about the importance of genre for audience pleasure, for example. You might want to look at p. 19.

Checkpoint 2

What is a narrowcast channel?

Exam question   answer: page 149

'The news automatically becomes the real world for the TV user and is not a substitute for reality, but is itself an immediate reality.' (Marshall McLuhan). Do we watch the news for information or for entertainment? Do you receive the reality of the information or the 'reality' of the entertainment?   (1 hour)

Cyberspace and identity

As progressively more of our media is being received through cyberspace, it is increasingly important to think about the nature of this space and about you, as audience, in this space. Roles are different in cyberspace – the audience is far more active and the institution must work in a different way. Ideologies are different and representations presented in different formats.

For these reasons, as well as looking at the media language in cyberspace in the previous section and looking at issues and debates raised by use of the Internet, for example, you must also think about the nature of 'cyberspace' and your identity and role within this.

What is cyberspace? ●●●

> *Cyberspace. A consensual hallucination experienced daily by billions of legitimate operators, in every nation, by children being taught mathematical concepts . . . A graphic representation of data abstracted from the banks of every computer in the human system. Unthinkable complexity. Lines of light ranged in the nonspace of the mind, clusters and constellations of data. Like city lights, receding . . .*
>
> Gibson, *Neuromancer* (1984)

Gibson is often quoted as having 'created' cyberspace in this novel. Cyberspace is defined as being the total interconnectedness of human beings through computers and telecommunication without regard to physical geography, as represented in Gibson's nightmare vision.

Since the end of the Cold War, new technological developments have supplanted political concerns in the scenarios of the science fiction film. Cyborgs and the simulated realities of media technology give us a perspective on our 'old' world, characterising it as inadequate or obsolete. Yet the cyborg in Hollywood remains a conventional representation of a 'baddie'.

In our everyday lives, a combination of reality and virtual reality manifests itself that effects a blurring of traditional boundaries. Cyberspace exists as a parallel space beside familiar real space, and we move comfortably between these two spaces, consuming media in both, as we choose. We are comfortable in both spaces and able to interact comfortably. We have become **'cyborgs'** with a merged identity which has physical elements and also cyber elements which are not present in the physical world. **Donna Hathaway** uses the concept of the cyborg to argue that hi-tech culture challenges and breaks down the binary oppositions that have always structured our perceptions. In this way she uses the concept of the cyborg to look at how gender, race or age is constructed differently in cyberspace and how different identities are constructed within the new media texts and by our consumption of these new media texts. In particular, she explores the relationships between the cyborg and feminist theories.

Action point

It is still fairly easy to get a copy of *Neuromancer* from Amazon or other book retailers and it can be a useful book to read.

Checkpoint 1

What is cyberspace?

Action point

Do you agree with Donna Hathaway's idea that we create an identity for ourselves in cyberspace which may be different from our physical identity?

Gender, race and ethnicity in cyberspace ●●●

Since our cyber identity is formed in different ways from our physical identity it does not depend on visual representations in the same way. Therefore representations of gender, race and ethnicity are different in cyberspace and there is (arguably) far more equality in cyberspace. Donna Hathaway argues that cyberspace allows non-sexist identities to be constructed, liberating women. However, many critics argue that new media texts still retain physical world stereotypical assumptions (think about the representations of Lara Croft in video games) and that this affects the way that cyberspace is constructed and controlled.

Regulation of cyberspace ●●●

There are a lot of issues which need to be addressed as convergence opens the possibility for all media to be delivered in cyberspace. For example, it is very easy for media producers to create any media text online as there are few restrictions and no censorship in cyberspace. The argument is that because the audience has 'chosen' to receive that text (by clicking on a link in most cases) they have made an informed choice. The problem comes in that there is no way of monitoring that 'choice' to warn an audience that it may be inappropriate, for example. Some critics would argue that this is important since it preserves freedom. Others argue that cyberspace should be regulated as with other media.

Action point

Do you know how you can protect yourself and your identity in cyberspace?

Information exchange in cyberspace ●●●

Another key development in cyberspace has been the growth in information sharing. At the most basic of levels this has taken the form of websites which have expanded beyond their original intentions to become significant texts and cultural features in their own right, such as eBay or Friends Reunited. However, similar technologies have been actively discouraged in other contexts, as when Napster was shut down for distributing music free of charge.

There are still peer-to-peer networks of this kind, such as Kazaa, and you may well use them. Do you think this is a good way of sharing files and information? Does it get abused in any way?

Checkpoint 2

What is eBay?

Institution and audience ●●●

Perhaps the most significant change in cyberspace is the active audience. Not only does the audience have to have actively selected a text in cyberspace but they will usually remain active. A game player cannot passively watch the game, but must interact. Because of this greater level of participation, it has now become quite usual for interactive media texts to have significant forum areas where game players can talk about the games and share their opinions.

Exam question   answer: page 149

Do you think there should be more censorship and control in cyberspace?

(1 hour)

Cult of the celebrity

Action point

How much celebrity-driven media do you consume? Do you consume these texts in spite of the celebrity focus or because of it?

Checkpoint 1

What is a 'sleb'?

Whereas once, the only real celebrities were seen as the Royal Family and the nobility, celebrity status is far more significant now. Not only do the Royal Family still have celebrity status but an enormous range of actors, singers, musicians, sportsmen, presenters, fashion designers, models, disc jockeys and other people are regarded (or regard themselves) as celebrities and increasingly have a particular niche and role within society.

Celebrity status

The celebrity is a person who is well known for being well known, a focus of gossip, of public opinion, of magazines, newspapers and the ephemeral images of movie and television screen.

Celebrity worship is a fairly recent phenomenon. While there were celebrity magazines such as *Hello* and the tabloid newspapers often had celebrity stories as part of their soft news, celebrities did not get anything like the attention they get now. If you look at a copy of *The Times* from even as recently as 1990, you will not find very much news about celebrities, and most certainly not on the front page. Today, though, if David Beckham or Posh change their perfume or shampoo, is a big story and papers such as *The Times* report it as if it were news we should all be concerned about.

Cult of the celebrity

We now have celebrities defined in particular genres, not just pop stars, etc. We have the cult of the celebrity gardener and the cult of the celebrity cook as well. We also have 'celebrities' who have appeared on reality TV game shows and become famous simply through being on these shows. They do a few interviews, release a yoga video and promote a couple of events and they have become famous. The press sometimes refer to these as C list (as opposed to 'true' A list) celebrities – or 'slebs'.

Representations of celebrity

As celebrity status becomes increasingly important in our society so have the representations of celebrity we receive. A celebrity can expect a paparazzi photographer to snap them anywhere at any time.

Glamorous, globally reported celebrity occasions such as the Oscars carry such weight that most of the stars are able to borrow outfits and jewellery for the occasion, knowing that the image and ideology of glamour they construct will also benefit the designers.

Paparazzi photographers

Photographers who follow – indeed, sometimes stalk – celebrities are known as paparazzi. The media market for their work – photographs of celebrities in both their public and private roles – is a lucrative one, offering high financial rewards for the most revealing images. While there are often stories about how a paparazzi photographer has

snatched photographs with a zoom lens without permission, many celebrities take care to work with the photographers, notifying them when they will be going out or leaving a restaurant, for example, or passing on information about another celebrity. Both sides in these agreements are aware of the power of these representations to make or break an image.

Celebrity endorsement

As we have mentioned in other sections, celebrity endorsement is very profitable. The star can usually choose products which will benefit the lifestyle ideology they have constructed and the products get the benefit of association. Many sporting stars and pop stars endorse products in this way and thereby enhance their own celebrity status since celebrity status depends on the amount of media coverage they receive and advertising campaigns extend that coverage.

Celebrity and the tabloid press

The tabloids rely on celebrities for soft news stories and, of course, we have a lot of celebrities in the entertainment scene – more than celebrities in politics, sports and business put together. Basically, we are not interested in our celebrities but in what happens to them – sickness, new hairdo, pregnancy, courtship, etc. This cult of the celebrity – more specifically, celebrity coverage – makes tabloids sell. The sex ingredient of this is obvious. So we get tabloids that carry entertainment news and gossip in more than half of their entire issue, with sexy photos designed to attract audiences who are not primarily interested in news.

Of course it is important to remember that this is a two-way relationship. The stars need the coverage – they have to be 'in the news' and they are desperate to remain famous, so they accept (and even encourage, as mentioned above) this coverage. This raises some complex questions when they complain that their privacy has been infringed by the media.

Celebrity magazines

Another growth area in the media is the celebrity magazine. Whereas once there was *Hello*! magazine with its glossy spreads for celebrity marriages and for celebrities to show off their glamorous lifestyles, now there are many more of these magazines, such as *OK*, *New* and *Heat*. As the marketplace has become more congested, the reporting has become more competitive and not all the features are so artificial. This has also led to a different relationship between the celebrities and the magazines, with the magazines being prepared to pay significant sums of money for exclusive coverage of events such as weddings or christenings.

Checkpoint 2

What is a paparazzi photographer?

Action point

Do you agree that the relationship between celebrities and the tabloid press is a two-way relationship?

Exam question   answer: pages 149–150

Looking at an edition of a national tabloid, in what ways are celebrity related stories presented in this edition? Account for these representations.   (1 hour)

Computer games

Checkpoint 1

What is a blockbuster game?

Action point

Why do you think games manufacturers are prepared to invest in producing the same game across different platforms?

Computer games are a growing and controversial media sector, driven not only by commercial developments but also by technological developments. Most contemporary games offer online gaming as part of the gaming experience and the technology available for playing these games is increasingly nearer to full virtual reality (VR). With these developments have come important media debates about violence in these games, the dangers of online gaming and ideologies constructed within these games.

Institutional issues

Video games are very big business. The amounts of money involved in designing and constructing a new game is at least equivalent to that invested in a new Hollywood blockbuster. Game play is increasingly complex and the games demand ever more sophisticated hardware. The release of a new blockbuster game, such as *Half Life 2* was given nearly as much promotion as a Hollywood blockbuster.

Games are available across a range of platforms and if you are researching this area, you should make sure that you have some ideas about these different platforms. Among the platforms you may wish to study are;

→ PC games
→ Xbox games
→ Playstation games (Playstation 1 and Playstation 2)
→ Gameboy games
→ Mobile phone/PDA games.

Most manufacturers will produce the same game across a range of platforms with slightly different game play to take account of the different platforms. But do they all assume the same male, young, white audience?

Representations and ideology

There are some issues that you need to think about while researching games. For example, many games across different genres depend on conflict. The player may be playing on their own (standalone) or they may be playing with other people using a network or across the Internet (multiplayer), but they are likely to be involved in conflict. There are exceptions, such as simulation games like *Sim City*, but the majority of games (as, indeed, the majority of conventional narrative) are based on conflict. Logically, therefore, by applying Lévi-Strauss' theories of binary oppositions, the conflict is going to be between the player and the 'enemy'. The representation of this enemy can be very significant.

→ A great many games assume 'alien' enemies and these can take several forms. However, they are usually humanoid to some degree though there is very rarely any sense of representation beyond a stereotypical 'alien' to be destroyed.
→ A lot of games are based on war scenarios, in which case the enemy is predefined. A Second World War game will assume Nazi enemies, for example. In this context it is interesting to note that *America's Army*

(a US military recruit training simulation) constructs representations of the enemy which are very Middle Eastern in appearance.

→ There are few female enemies in games. Female characters are often constructed within the narrative but they are generally a helper for the protagonist rather than protagonists themselves. *Tomb Raider* was an exception here, yet still encouraged male game play – how?

Violence in computer games ●●●

There are many studies of the violence in computer games and how this may affect players – especially children. Many games require players to destroy opponents of some kind and the game play can be very graphic.

One way the industry has tried to address this is by regulating games in a similar way to films. Games are given a rating, e.g. 3+, 11–15 and 18+. This still leaves a lot of issues to think about:

→ Is the rating system effective? Can young children get access to games for older age groups?
→ Do gamers know the difference between games and real life?
→ Are violent games marketed differently from less violent games such as simulations like *Sim City*?
→ Gaming is an active pastime – in other words, you have to actively play the game, not passively sit and watch it. Does this make the levels of violence any less acceptable?
→ Do boys and girls play games differently? Do they want to play different games?

Consider this statement from a member of the BBFC computer game classification team:

A game which, for example, rewards anti-social behaviour without consequence is likely to be placed in a more restrictive category than one which gives free rein to aggressive impulses – Carmageddon *being a prime example. The concern here is mostly to do with possible effects on impressionable children and younger people, coupled to a psychological model which suggests that rewarding anti-social behaviour at an early stage of development is likely to have a lasting impact upon them in later life and, consequently, on the greater society at large.*

Online gaming ●●●

The role of the player in an online game is more complex – the player takes on the role of a character, usually within a team, to play out a scenario. There is not usually a predetermined resolution to the narrative as that will depend on the players. Environments may change, situations will change and the character may evolve in different ways. Does this mean the player is in control of the violence or ideologies in the game? How important is communication?

Checkpoint 2

Are computer games classified?

Exam question   answer: page 150

Compare and contrast the representations and ideologies operating in an adventure game and a simulation game of your choice. In what ways do they indicate the target audience for the games?   (1 hour)

Popular music

Music remains fundamental to our culture. From film music, muzak in shopping centres, polyphonic ring tones on a mobile phone, MP3 players and iPods, we are surrounded by music all day. As the technology improves, so does our level of access in many cases, whether we want it to or not!

The music industry is a central part of this music culture and is constantly changing in order to maintain and develop audiences. Marketing becomes ever more important and more and more music artists are realising that they must move from 'band to brand' if they are to survive.

Development of pop music ●●●

Music has always been closely linked to technological development. Historically, music was all live performance by the musicians or singers. As silent films took off, so there was a pianist playing live in the cinema to create mood and atmosphere for the audience. Even at this stage, the music was culturally defined (a lone pianist in full evening dress playing on a raised area) and gave status to what was otherwise seen as more lowbrow than theatre.

With the invention of the mass-produced record came pop music – and the most significant thing about this was the invention of the 'teenager'. The teenager did not exist until created as a cultural label to define (at that time) a particular kind of adolescent rebellion. The identity of the teenager was defined first by the fact that they listened to pop music and secondly by the clothes they wore to show that they belonged to this music culture.

As pop music became more popular, so subcultural groups developed, differentiated from each other by style of music and also style of dress. Hence the pitched battles between Mods and Rockers in the 1960s, for example. Today the 'music scene' is considerably broader and there are many different ways of receiving music from countless different sources. Each of these musical genres still has an associated subcultural group and constructed image.

New technologies and pop music ●●●

Technological development continues to influence the ways in which music is received and therefore affects the cultural structures around this consumption. For example:

→ Technological developments have enabled far more artists to reach a wider audience. Whereas once, recording studios were hallowed places with a four-track mixer considered sophisticated, now many artists have top quality recording studios in their bedrooms and are easily able to produce their own music. In many cases they are also able to distribute it without needing record companies at all. This is also why commercial artists use image and brand as much as the music itself to promote themselves, knowing that it is this visual sophistication which sets them apart from the amateurs now.

Checkpoint 1

What is a subcultural group?

Watch out

It is very difficult now to separate music and image for most music artists so you must remember to research both together.

→ Napster generated many debates about the morality of sharing music on the Internet for others to download. Napster, when it was first operating, was the first of the peer-to-peer networks which allowed people to share files easily in this way. It was soon shut down by the record companies who accused Napster of losing them money, insisting that it was depriving them of audiences who might buy the music.

→ As a result of realising the popularity of downloading music from the Net, many of the record companies now have this facility and indeed many bands will make some of their music available free on the Net, knowing that this will gain them additional audiences.

→ Technology has also changed the way we consume our music – MP3 players and iPODs are commonplace and almost all artists produce videos to go with single releases as a matter of course. 24-hour music television then promotes these videos (MTV, etc) and encourages audiences to identify with the music and the video.

Representation and music culture

Representations in relation to popular music culture break down into two distinct groups:

1. Representations that artists construct to represent them, their music and ideologies and create a clear brand for the audience. This image will probably change to reflect each new release.
2. Representations which are therefore copied by audience members who wish to demonstrate their association with a particular style of music and particular artists.

Genre and music culture

Genre is fundamental to popular music. Genre is probably a more significant concept in relation to pop music than even to film! Some genre categorisations are fluid but others are rigidly defined and artists and audience alike seek to maintain the purity of that genre, arguing passionately about the music and accusing others of selling out if they move outside the genre conventions and expectations.

Of course, genre is also very useful to retailers, because they know that the audience tend to identify with particular genres and buy music primarily from that one genre. This is where popular music is different from other media texts and is almost certainly to do with the way that music artists manage to create such an intense ideology with their careful construction of brand for their band.

Action point

Do you think that you should be able to share music free in this way or do you agree that the artists should be paid for their music?

Examiner's secrets

Remember that if you are researching music culture for Media Studies you must explore representations, ideologies and institutional issues in depth, not simply focus on the music as that is not directly associated with the key concepts.

Checkpoint 2

What are the two stages of representation in relation to music culture?

Action point

How many different genres of popular music can you identify in your local music store?

Exam question   answer: page 150

With reference to the promotional materials for the new release from a music artist you know well, how are representation and genre used to attract a target audience?   (1 hour)

Crime and violence

Crime and violence in the media are a broad area of study, involving study of the representation of crime and criminals in the news, issues and debates arising from this portrayal, such as moral panics, and also the representations, ideologies and generic conventions employed in crime-related drama on television.

Crime and police series on television

Police series and crime series are always popular on television. *The Bill* remains one of the most successful programmes on ITV and has as loyal audience as any major soap opera. Crime is now regarded as a key genre for television and, as such, has its own codes and conventions:

→ Most of the action in a crime series is based on binary oppositions.
→ There is usually a chase sequence – often a car chase.
→ Mise-en-scène is often dirty and cramped.
→ There is a lot of emphasis on 'tools of the trade', e.g. guns. In the latest high-tech crime series these are often expensive, sophisticated and, at times, beyond what the current technology can achieve.
→ The lead character is usually a troubled person with a complex past and personal tragedies.
→ The hero usually has a sidekick.

Links

This debate is closely linked to effects theory so you may also want to look at pp. 86–87.

Violence in the media

Cultivation theory suggests that we are increasingly desensitised to violence in the media and the evidence would suggest that that is the case. Our tolerance for violence has changed over time and we are accustomed to receiving very graphic news coverage 'live' from around the globe now which we would not have tolerated in any context some years ago. Lyrics which would have ensured an instant ban for a rap song some years ago are accepted as normal now.

There are various arguments about the influence this has on our lives which you should be aware of in relation to this topic area:

Checkpoint 1

What is catharsis?

→ **Influences on behaviour** – Many psychologists and others report that children, in particular, are influenced by watching violence on television and become significantly more aggressive.
→ **Constructed realities** – By watching significant quantities of violence on television we can start to believe that it is extensive and lose perspective on reality.
→ **Catharsis** – Some theorists argue that experiencing violence second-hand through the media can enable us to release violent feelings and actually be more peaceful.

Moral panics

Moral panics start when the media are placed in a position of becoming an agent of social control. The panic which is generated by a story of this kind (e.g. about asylum seekers or devil worshippers) relates to the

theory of deviancy. As you know, most news depends on binary oppositions to establish a story (good/bad) so any story which represents a group or individual as dangerous will end up classifying them as 'bad' and therefore what we call 'deviant'. The problem is that a moral panic can cause more deviancy and does not resolve situations.

There are three stages to the creation of a moral panic:

1. An event occurs which generates media attention (e.g. a murder or a sexual abuse case). If there is an opportunity for the media organisation to cover this in detail, it immediately gives the event status for the audience and causes anxiety.
2. Media organisations relate this one event to wider social issues (e.g. a story about an asylum seeker generates substantial coverage about the 'problems' and 'dangers' of asylum seekers, or a story about a paedophile leads to coverage about protecting children from paedophiles in general).
3. Media coverage increases, social pressure forces government action of some kind and the public feel vindicated. The *News of the World* created a moral panic of this type some years ago with their 'outing' of paedophiles living in the community.

Children and violence on television

Assumptions about the dangers of exposing children to violence in the media are commonplace. One good example is the Newson Report (1995) following the Jamie Bulger case, which concluded that violence in the media led directly to real-life violence and that this situation would continue to get worse. However, critics of the effects model point out that this research has certain flaws. For example, the research categorised all types of violence from children's cartoons, news programmes and hardcore pornography as the same and insisted that all would impact on children in the same way.

David Gauntlett, in particular, has identified ten things wrong with the effects model – especially in relation to children and violence in the media. This is a useful critique for many of your essays relating to audience theory. He argues that the effects model makes many incorrect assumptions, such as:

→ assuming children cannot cope with violence in the media in any form
→ not differentiating between types of media
→ assuming a passive audience and a single meaning to be received from media texts.

Checkpoint 2

What are the three stages to a moral panic?

Check the net

You can read David Gauntlett's report about this at http://www.newmediastudies.com/effects1.htm

Exam question   answer: page 150

Do you think there should be more controls over violence in the media?

(1 hour)

Politics and propaganda

Checkpoint 1

What is a sound bite?

Watch out!

Politics is increasingly becoming media-driven so you should take time to research the ways in which politicians use the media carefully.

Checkpoint 2

What do we mean by 'celebrity endorsement' here?

Propaganda has always been an important area of study for Media Studies. From the Frankfurt School analysis of Nazi propaganda to the way the former Soviet Union used the media as a means to control the masses, there has been much to study. In more democratic societies, however, the media have not been used as a control tool in quite the same way. For example, in Britain, we do not have a state news agency which can only disseminate approved news items, or a state television station which can only broadcast approved programmes. Yet, like so many other areas of contemporary society, such as sport, the government has increasingly come to realise the power of the media and to use the media to good effect.

The background

Until fairly recently, the relationship between the media and the government was fairly simplistic. While there were political correspondents for each of the major news organisations, who knew what was happening in government and reported significant stories to the public, and press officers for each of the major political parties, for example, it was not a close relationship. However, politicians have become increasingly aware of the importance of a 'good press'. For example:

→ Politicians make a point now of speaking in sound bites, i.e. short, snappy phrases which are easy to edit into a news bulletin.

→ Photo-opportunities are increasingly important for politicians to promote their positions. Whereas once it was traditional for politicians to kiss babies when canvassing for election, now, politicians will ensure that they are regularly featured in their local newspaper, for example, for constant coverage.

→ Newspaper owners (e.g. Rupert Murdoch) are courted by politicians who know how much influence the newspapers can have on elections and public opinion.

→ There are increasing numbers of political commentary programmes on television and more politicians are prepared to take part, e.g. in *Question Time* or on *Newsnight*, to persuade the public to back them.

→ All major political parties are entitled to show party political broadcasts, especially at election time. These broadcasts must be shown by each of the mainstream television channels on a particular day (although no longer necessarily at the same time). Again, early broadcasts were simple campaign messages but now are far more sophisticated, with very high production values and often spending more time criticising other parties than announcing policy.

→ Celebrity endorsement is becoming increasingly important, especially at election time. Politicians hope that the public will be influenced to copy these celebrities.

→ News is now broadcast and reported live or very soon after events, so politicians have had to learn to be skilled interviewees.

Spin doctors ●●●

Spin doctoring is another concept which began in the US. It means to put a 'spin' on a story – in other words, to present it in a particular way, to put a particular meaning on it. So, for example, a political spin doctor would present a story about increasing taxes not by looking in detail at how much extra money people would have to give in taxes but the benefits this would generate in terms of hospitals, education, etc.

Their main job is to deal with journalists and to explain the party's or government's position on an issue; they will be in constant contact with journalists. The more senior ones will be key advisers on electoral strategy and have an input into general government policy to ensure this positive stance. Spin doctoring is the art of putting the best possible gloss, the best possible presentation on a political speech, or on a political event. And is very important in today's fast-moving news agenda **to put the best interpretation on what has just happened**.

Political control of the media ●●●

There are several ways in which politicians try to control the media in contemporary society:

→ By influencing the news agenda (e.g. by releasing a big story to try to 'kill' a negative story by limiting space for coverage).
→ Influencing the way a story is presented (e.g. using spin doctors).
→ Using the media to test public opinion on possible changes (e.g. 'leaking' stories about changing the school leaving age to judge public response before announcing new policies).
→ Using the media to discredit party members who speak out against party policy, for example, and putting pressure on others not to speak out as well.

Politics, war and the media

During a time of war, the media are even more important to the government. Not only must the public be reassured at home but the media are a powerful weapon to convince the opposition that they are losing. The government may use D notices to prevent the media reporting certain events and will usually provide many press briefings for the media to report appropriately.

Checkpoint 3

What is a spin doctor?

Check the net

The *Guardian* newspaper has a large archive collection relating the the war in Iraq, which can be accessed at http://media.guardian.co.uk/iraqandthemedia/0,12823,883261,00.html. This includes articles about the way journalists were controlled in Iraq and the way the government sought to control the media during the war.

Exam question   answer: page 150–151

Choose a party political broadcast from a major political organisation (such as the Labour Party, Conservative Party or Green Party) and give a detailed deconstruction of what ideologies are promoted in this text. (You can usually download recent broadcasts from their party websites.)   (1 hour)

Sport and the media

Not only do all the terrestrial TV channels offer sports programming, there are increasing numbers of specialised satellite and cable sports channels. Not only are there many niche sporting magazines but all newspapers carry a sports section with very distinctive and common codes and conventions. Not only do all news broadcasts end with sports reports but there are even dedicated sports radio stations. And as new media increase the reach of sports broadcasters and the technology continues to develop, there will always be sporting audiences. The ideologies and representations relating to 'team' and 'sport' are an important area of study.

Sport and the media ●●●

There are many different ways that audiences receive sports coverage:

→ The back pages of all newspapers are dedicated to sport (in priority order, with primary sport, i.e. football/rugby on the very back page and minority sports further in).

→ All news programmes finish with a sports section – usually only dedicated to mass audience sports such as rugby and football.

→ Terrestrial television provides dedicated sports coverage, including substantial coverage on Saturdays.

→ Most radio stations provide some sports coverage and some are primarily sports-centered such as BBC Radio 5.

→ Subscription-only premium sports channels such as the Sky Sports channels have some of the largest subscription audiences.

→ Audiences are prepared to pay per view for many premium sporting events, such as global broadcasts of key boxing matches.

→ Part of the PSB requirement for Channel 4 is to provide minority sports coverage.

→ Sports headlines can be sent directly to your mobile phone.

→ Major sporting events are given great prominence – pushing normal programming out of the way on terrestrial television, for example.

Sports audiences

Sports audiences are predominantly male, with about three-quarters of men regularly watching sport on television and/or reading about sport in the news. The major sports in the UK are football and rugby and by far the majority of sports broadcasting and reporting is related to these two sports. Some media commentators suggest that one reason for this is that one of the functions of sport in society is the construction and validation of masculinity and male superiority, both by playing and consuming sport. The way in which these different sports are promoted and the amount and type of media coverage granted to them would suggest that these are important features of our culture.

Checkpoint 1

Why is it usually football which is shown on the back page of a newspaper?

Don't forget

Sports reporting is one of the most profitable forms of media – there will always be large audiences keen to receive sporting events.

Watch out

Remember that there are many more ways to receive sporting media than simply the newspaper back pages, news items and *Grandstand*!

Representation

How the media portray men and women in sports is another important debate. Many critics argue that the media reinforce traditional ideologies, consciously or unconsciously emphasising male supremacy and reducing women to sexualised, subordinate and maternal roles in the language of sports commentators and journalists, the composition of sports photographs, public discussions of sport funding policies and so on. Women's sport is not given anything like as much attention as male sport and, at times, the coverage of women's football or women's boxing, for example, can be very negative.

Ideologies

As mentioned above, the representations of sportsmen which are generally used glorify them whereas the representations of women playing sport are often far less positive. The ideologies and values of sport are used throughout society – we talk about 'winners' and 'losers', 'playing the game' or 'losing the toss', for example, and many newspaper institutions admit that it is the sports coverage that helps them to maintain their readerships.

Institution

The media and sport have become inseparable. This can also involve complex relationships with advertisers, who may **sponsor** a team or a competition in order to promote their brand, for example, by 'branding' players or equipment. Many sports institutions sign up celebrity sports players to wear their clothing ranges or use their equipment, knowing that media exposure will promote their products because the audience will aspire to the ideology constructed by the celebrity.

New technologies

New technologies offer the world of sport a range of obvious advantages which should be exploited. For example:

→ Streaming media is allowing increased global exposure for minority sports via the Internet.
→ Weblogs, websites and headline messages to mobile phones, for example, can allow far greater audience interaction with sporting events, especially those taking place over a period of time.
→ Wider channel choice is allowing an expansion in sports broadcasting on television and radio.
→ Interactive televison is increasing audiences for sporting events, e.g. during Wimbledon, the ability to watch more than one match at a time is very attractive to audiences.

Watch out!

As sports broadcasting continues to increase, so sports personalities are beginning to take care with their visual representations, just as with politicians.

Checkpoint 2

What is the advantage to an advertiser if they sponsor a sporting event?

Exam question   answer: page 151

Watch a sports broadcast such as *Grandstand* or the broadcast of a sporting event such as a Premier League football match. What ideologies are constructed in the broadcast? What representations are used to establish these ideologies?   (1 hour)

Women and film

Women and film as an area of study is a popular and important topic area in Media Studies. It can link with study of feminist theory, representation studies, theories of audience, institutional studies and historical studies of film. The film industry, as with so many media industries, has historically tended to be primarily male in terms of production and, to some extent, consumption. Women and film as a topic area can encompass studies of individual film directors or actresses, representation of women in film, female audiences and 'female' films, among others. If you are researching this topic area in depth you should check with your teacher about exactly what your research area will be.

Directors and actresses

There are many directors and actresses whom you may study in this research area, who have made significant contributions to the development of film, made outstanding films or been significant in influencing the form and shape of individual films or even genres. For example, you might be studying:

→ Jane Campion
→ Kathryn Bigelow
→ Marilyn Monroe
→ Sigourney Weaver
→ Cate Blanchett
→ Halle Berry.

With any of these women as your topic source, your obvious starting point is to watch all their films and explore chronological changes as well as identifying significant moments in these films. You can also research them in textbooks, journals and on the Internet but remember that your starting point should always be the films and you should always make sure that you have specific examples you can use from these films to support points you are making.

Representation of women in film

There are many ways that you can explore representation of women in film and how it has changed over time. From the Golden Age Hollywood studios, with their very specific requirements about the appearance of female stars, through to the expectations placed on contemporary actresses to conform to specific ideologies about appearance, this has always been an important area of study. Aspects include:

→ A study of how female image was constructed in film noir (the heroine and the femme fatale) and how that has changed with contemporary neo-noir films.
→ The male gaze in relation to contemporary Hollywood films.
→ A comparison of the representation of women in 'chick flicks' and 'teen movies', with a focus on specific films.
→ Particular actresses as archetype (for example, Meg Ryan who now become an archetype in society as a whole).

Watch out!

Remember that, whatever area you decide to research, it is very important that you choose specific films to study so that you can use detailed examples to support the points you make.

Checkpoint 1

What is an archetype?

146

- → The use of stereotypes in a particular genre, e.g. in horror movies.
- → A comparison of representations of women in different national cultures, e.g. Hollywood and Bollywood.
- → The changing image of a particular actress to conform to specific expectations (e.g. Renee Zellweger – especially in relation to the pressure she felt because she had to put on weight for the *Bridget Jones* films).

Female films

As female audiences continue to grow and the way that films are consumed changes (DVD, cable channels and a much broader global market, for example), new genres of film are developing, such as the chick flick and teen movies which are clearly aimed at women, not men. Some characteristics of these genres are:

- → A combination of comedy and other genres or a very emotive, 'tearjerker' ideology.
- → Strong female representations, usually across a range of ages, and not always glamorous (as opposed to typical action adventure where actresses are expected to be young and very glamorous).
- → Even teen chick flicks tend to have an obvious moral conclusion.
- → A range of male representations (although these are usually subservient to the women).
- → Less explicit sex than in other genres but more focus on emotions and love.

The relationship of women and film

Media institutions use a range of strategies to ensure their products are successful and this has led to intense scrutiny of women in film, for example.

- → Film institutions use actresses in particular ways to sell films, especially action/adventure films. Many actresses complain that they are treated as 'meat' by the institutions and given very little respect. For example, a female star will rarely headline a film above a male star.
- → The images constructed for these female stars are promoted very heavily throughout the media and put immense pressure on women around the world to look like that. This has led to ever-increasing demand for plastic surgery, etc, as women attempt to conform to these ideologies – yet many actresses themselves insist these are false, fabricated representations.
- → Audiences are presented with these ideologies and therefore these representations are received – the debates are about the extent to which this is a passive reception or whether audiences are able to negotiate meaning from these representations. Of course for some social groups these are very oppositional readings!

Checkpoint 2

What is a chick flick?

Action point

Why do you think that chick flicks tend to have a moral?

Links

You may want to relate this to feminist theory – for example, Mulvey's theory, by looking back at p. 108.

Exam question   answer: page 151

Compare and contrast the representations of women in a current Hollywood blockbuster and a current Bollywood blockbuster. Do you think these are legitimate representations?   (1 hour)

Answers
Media debates

Documentary forms

Checkpoints

1 Actuality filming is unedited filming of events, such as early works from the Lumière brothers.
2 Lifestyle programmes are generally seen as aspirational because they all focus on 'lifestyle improvements', i.e. implying that if you follow their advice you would improve your life.

Exam question

Nick Broomfield and Michael Moore are both well-known documentary makers who believe passionately in the role and function of documentary to act as a record of the truth, not merely a representation of reality. Both look for subjects which they feel are important and prefer to create documentary films rather than TV documentaries because they prefer the freedom that film offers them. They produce documentaries in the direct cinema school of documentary production, allowing subjects to speak for themselves, and often appear in the frame, for example, asking questions – not as 'voice of God'. In this way, it can be seen that both of them are seeking to remain true to the ideal of documentary making, using the most traditional forms of documentary construction to create texts which are deliberately subversive and challenging. It is this reliance on the most pure form of documentary production which enables them to challenge their subjects, unlike more modern documentaries which make use of more varied cinematic form but without seeking to subvert or challenge. So, your answer will probably look in detail at the way they construct this subversive or challenging ideology through representation, framing, camera and editing and how this uses conventional form to establish the challenge.

Censorship

Checkpoints

1 The PCC is a self-regulatory body, which means it is run by the press organisations and these organisations can also choose whether they wish to belong to the PCC or not.
2 The ratings system for films, etc, only applies to watching films in a cinema or renting/buying them. There is nothing to stop you watching the film at home as long as you did not buy or rent it.

Exam question

Desensitisation is an important concept within Media Studies because there is a lot of evidence that we are becoming desensitised – a term used in cultivation theory to explain the way repeated exposure to levels of violence or sex makes them more normal and thus alters our expectations of what is abnormal. There is plenty of evidence to show that we are becoming desensitised, for example, the number of films which are finally released after having been previously banned or having scenes cut. Invariably, our reaction is: 'Is that it?' – yet at the time the material was considered unsuitable for release. The question therefore is asking how far this might go. Of course, an answer to this question could become a list of 'what might be suitable one day', but this would be unlikely to score highly. However, an answer which considered the role of the BBFC in terms of effects theory or of whether it is only the representations and visual imagery which change or the ideologies also, would be far stronger. This might lead to discussions about changing media exposure, with increased access to media on demand and extensive media choice. These might suggest that the BBFC is indeed redundant, given the sheer volume of other media available which negates the value of controlling film and video release.

Media ownership

Checkpoints

1 A conglomerate is a collection of companies owned by a single organisation. Most of the large media organisations are conglomerates of vertically and/or horizontally integrated companies.
2 Diversification refers to the way these organisations increasingly move into new sectors of the industry.
3 Convergence refers to the way that the large media organisations increasingly dominate across all media sectors, bringing a range of media production into a single organisation.

Exam question

This is a relatively straightforward question, depending on the level of your research before attempting it. However, this is not to say that your answer should simply be a series of facts about the ownership of the organisation. The question asks about the ownership profile so stronger answers would consider how the organisation is run; the way the organisation structures itself (horizontally or vertically integrated, for example); the role of the organisation within the relevant industry sectors, and the ideologies which the organisation constructs and maintains. The stronger answers therefore are likely to be centring on establishing why the ownership profile is as it is, and how this reflects the context in which the organisation operates, not just identifying what the profile is.

Globalisation

Checkpoints

1 Cultural homogenisation is the way that our cultures are increasingly convergent because of the shaping of global media – we are all tending to take on US cultural values and ideologies.
2 Cultural imperialism is related to this, as critics argue that this US domination of global media is almost a form of imperialism.
3 The global village is McLuhan's definition of the way we receive news and information from around the globe instantly – as if we were all living in the same small village.

Exam question

Your answer here should start with an identification of the current market reach of Disney across different media sectors and in different contexts. Having identified what products Disney produce, you should go on to consider whether this can be called a brand. There are two ways you might respond to this – you might argue that it is a brand, since all the products are readily identified as Disney and there are brand expectations quickly established – you might also argue that it is not a brand since the ideology is so strong that it is more than a brand. In this case you might well extend your analysis to consider contexts such as Disneyland and how they establish and maintain the 'magic'.

To conclude your essay you should answer the question directly as to whether this is cultural imperialism at work. You don't need to worry too much about explaining what cultural imperialism is but you will need to argue whether Disney simply has a very strong product brand or whether there is a form of cultural imperialism working here – and if so, what ideologies permeate it.

New media and the information society

Checkpoints

1 The information age refers to the fact that our society now depends on information rather than production. More and more information is available and we are often described as 'information rich and time poor'.
2 A narrowcast channel is a cable or satellite channel which targets a very specific audience – in the same way that a magazine may target a particular hobbyist niche market.

Exam question

Although it might seem that there are two questions here, in reality this is just one question about why audiences choose to watch/read/listen to the news. Is it for entertainment, just as with soap operas or horror movies, or is it to keep up with world events? This is a question which needs to focus on specific texts for a detailed answer and is probably best suited to an examination of television news, since the relationship between information and entertainment is probably closest for television news. Very strong answers might extend this to consider Baudrillard's assertions about the lack of reality in reporting of the Iraq war on television, constructing a reality of glory, goodies and baddies and spectacular weapons, deployed by heroes, which bore little relation to the reality of events in Iraq. Answers therefore will probably centre around the balance in news reporting between reporting the reality but in a form which suits an entertainment medium, or entertainment which simply derives from real events. It may be that your answer will conclude that the balance is different for different channels or different audiences, for example, but this, in its turn, would lead to further discussions about the role and function of news reporting and the wider contexts of production and broadcast.

Cyberspace and identity

Checkpoints

1 Cyberspace is the 'place' where we interact via technology such as the Internet, chat rooms, Messenger, etc. It does not have any physical presence at all.
2 eBay is a website which puts people who have items to sell in contact with people who want to buy them. It is probably one of the most well-known websites in the world. There are some very good second-hand bargains to be had on eBay, but there are also stories of people losing a lot of money.

Exam question

This is another wide-ranging and complex question which needs a clear focus if it is to be answered in sufficient detail. There are many issues to consider here, such as the ease with which identities can be constructed in chat rooms, for example, the constant streams of viruses and spam which continue to be produced, and the ease with which websites can be plagiarised.

This is a complex area and, as with so many topics, you can argue in different ways. There are those, for example, who argue that it is a good thing that there are relatively few constraints in cyberspace and that this freedom is worth occasional infringements. Others will argue that cyberspace needs legislation, for example, to protect children from surfing unsuitable websites such as porn sites.

Of course, you can see that this is basically another form of the effects versus pluralist debate and answers are likely to build on this debate. Stronger answers will do this with use of specific examples and case studies, not simply addressing the title in general terms without evidence and detail to support points made.

Cult of the celebrity

Checkpoints

1 A 'sleb' is a dismissive term coined by the tabloid press to describe someone who is trying to be a celebrity but is not really well known or successful enough. It is often used of those who are trying too hard to be celebrities.
2 A paparazzi photographer earns their living by following celebrities (including royals) and taking photographs of them which they can then sell to newspapers and magazines. Usually these pictures are taken without the celebrities' knowledge.

Exam question

This is another question which is grounded in detailed textual analysis. The content of an analysis here will obviously depend on the chosen tabloid and the chosen story but should be grounded in the key concepts. Therefore a secure answer would look at the representations used in relation both to the ideology and image of the celebrity being represented and would also contextualise this in terms of the institutional context of the tabloid. Issues to debate here

would revolve around the relationship between celebrities and the tabloid press – is the celebrity actively courting publicity (of whatever kind) in order to enhance their profile, for example? In this case you might also reflect on why the tabloid has responded this way. Or is the tabloid instigating the article perhaps representing the celebrity in a negative way for some reason? Theoretical issues raised might well be focused on audience uses and gratifications from this publication; how the audience constructs celebrities to conform to certain stereotypes, and the ways in which celebrities must then work within those parameters.

Computer games

Checkpoints

1 A blockbuster game is one with high production values and a large promotional budget, just as with a blockbuster film.
2 Computer games are classified in a similar way to films and it is also done by a division of the BBFC.

Exam question

This is a typical deconstruction-type question where you have to identify the target audience for your two chosen games, analyse the representations and ideologies constructed (not just through visual signs but also sound, packaging, wider contexts, e.g. film links, etc) and then compare the way these relate to their specific target audience. The target audience for the two games is possibly going to be similar demographically although the psychographic profiling is likely to be different – however, this will depend on the particular games which you have chosen to analyse.

Popular music

Checkpoints

1 Subcultural groups are those that form within a larger cultural group. Subcultural musically related groups in this country are usually defined by the genre of music they like and this influences their visual representations as well.
2 The two stages of representation in relation to music culture are firstly the representations constructed by the artists and secondly the representations constructed by those who wish to demonstrate their association with this type of music culture.

Exam question

Clearly, a response to this question will be significantly different, depending on the choice of artist. However, these differences will be in details and examples rather than in structure. A strong answer to this question will approach the question through the ideology which is established by these promotional materials.

Your answer should cover not only how the main artists are represented but also how these representations are employed to create meaning. For example, Madonna remodels herself for each new release and establishes a

very clear ideology and context for each new album. As part of this representation of her new image she also builds a range of representations around it which are echoed throughout the materials in various ways.

Genre is also an important area for music releases – given that music is often identified by genre before content, it is important that the genre is communicated immediately by the promotional materials. This is likely to be done not only by the representations but also by use of colour and font, for example, or vocabulary used in puffs, etc. It might also be reflected in the form of the materials.

Crime and violence

Checkpoints

1 Catharsis is emotional release. From classical theatre onwards, powerful drama has enabled us to experience intense emotions and therefore release our own pent-up emotions in a safe way – the cathartic effect.
2 The three stages to a moral panic are the event, the media reporting of it and the saturation coverage which generates intense pressure.

Exam question

This is a more general question, of the sort which you might get at A2 level. As with other questions of this sort, it is important to start by thinking about what texts you can use as evidence to support your argument. An essay which does not have a detailed focus will be less successful because it will be too vague. You cannot hope to cover this entire subject in an exam essay, so the examiners will be expecting you to apply your knowledge in a specific, detailed analysis. So, for example, you might write about desensitisation or effects theory, or perhaps about how global news coverage has increasingly become more violent, and in each case you would use particular texts to support your argument. Strong answers will link the theory and the texts very closely, demonstrating that you understand the theory and that you have thought about the texts that you are writing about.

Politics and propaganda

Checkpoints

1 A sound bite is a short phrase which the politician uses deliberately, knowing it is a good length to be included in a news broadcast.
2 Especially at election time, politicians are very keen to persuade celebrities to be seen as backing their political party, knowing this may influence voters.
3 A spin doctor is responsible for putting the best possible 'spin', or representation, on any story relating to the political party. This may mean putting a positive spin on bad news, for example.

Exam question

As with many of the other questions in this section, this question asks you to do a detailed textual analysis of your

chosen party political broadcast, in relation to the key concepts. Clearly, an answer about political media of this kind is probably going to start with a focus on ideology, since that dominates the structure of a party political broadcast. A useful way of approaching texts of this kind can be through narrative theory – there are often very explicit representations of 'goodies' and 'baddies', for example, and use of enigma codes to entice an audience to buy into the ideology on offer.

Sport and the media

Checkpoints

1 The back page of the newspaper is where many male members of the audience start reading, so the most important sporting events are always placed on this page, usually with a large central image.

2 If an advertiser sponsors a sporting event, they will not only be able to use their logo throughout all the promotional materials but also during the sporting event and they will ensure further publicity by presenting the prizes to winners, for example. They usually choose a sport which gives them access to a suitable target audience.

Exam questions

Again, this question is asking you to analyse a particular text with close reference to the key concepts. Strong answers here will be able to move beyond a basic textual analysis, however, to consider how sporting ideologies are promoted through sports broadcasting, the use of sporting celebrities to attract audiences and add status to the programme, and the general competitive atmosphere which is central to audience expectations. Wider ideologies are often presented through the combination of live action and commentators who are given status in various ways such as costume, mise-en-scène and editing. The commentators shape the audience responses to the sporting action and the action is mediated by the programme context which leads to important debates about the 'reality' (or perhaps 'hyperreality') of the events, for example, and the power of the media to control sport and shape our expectations through these representations.

Women and film

Checkpoints

1 An archetype is the model which is copied by many others. Meg Ryan, for example, is the archetypal dizzy blonde in film and there have been many others since.

2 A chick flick is a film which is targeted at a female audience; it is expressly 'female' in approach and hence less likely to appeal to a male audience.

Exam question

A response to this question is likely to start with a focus on stereotypes. Each of these contexts carries particular audience expectations about stereotypes and cultural expectations and clearly this is important to the answer. A strong answer will make use of the two primary texts in detail and may well reference other texts to support points made. However, stronger answers will also go on to consider these representations in greater detail. For example, the visual representations of women in most Bollywood films are very conventional, whereas the ideologies operating often show women as far more independent, resilient and powerful. In typical Hollywood movies, the female stars may be presenting characters who appear more independent and powerful but are generally subjugated to the male characters and subject to the male gaze in a very direct way.

You should also consider wider cultural contexts here and how these shape the expectations of the representations. Think, too, about the extent to which these representations can indeed be considered stereotypical or whether they are accurate. (This may lead to debates about the relationship between the audience and the text in terms of pluralist theory or uses and gratifications, for example.)

Lastly, you are asked to evaluate whether these representations are legitimate, which is another complex area. Legitimate might be appropriate for the filmatic context, or it might imply cultural legitimacy or genre legitimacies. The purpose of this second part of this question therefore is to open it out to these debates around contexts and ideologies rather than remaining centred on stereotypes.

Media debates
Revision checklist

1	Understand the basic context within which you could do some specialist research about women and film	Confident	Not confident. **Revise** pages 146–147
2	Understand the basic relationship between sport and media and identify a case study to research further	Confident	Not confident. **Revise** pages 144–145
3	Understand the development of pop music and its relationship with media and identify a case study to research further	Confident	Not confident. **Revise** pages 138–139
4	Understand the importance of media for politics and the ways politics uses the media	Confident	Not confident. **Revise** pages 142–143
5	Understand about the information society and how this affects our cultural context of consumption	Confident	Not confident. **Revise** pages 130–131
6	Understand about vertical and horizontal integration, synergy and convergence in relation to media organisations	Confident	Not confident. **Revise** pages 126–127
7	Understand how globalisation is affecting the context of production and consumption for media texts and understand about cultural imperialism	Confident	Not confident. **Revise** pages 128–129
8	Understand the basic issues in the debate about the relationship between computer games and violence and identify a case study to research further	Confident	Not confident. **Revise** pages 136–137
9	Understand the codes and conventions of varying genres of documentary and the way documentaries are structured	Confident	Not confident. **Revise** pages 122–123
10	Understand how identify is constructed in cyberspace and how this affects media texts in cyberspace	Confident	Not confident. **Revise** pages 132–133
11	Understand debates relating to violence in the media and identify a case study to research further	Confident	Not confident . **Revise** pages 140–141
12	Know the basic forms of censorship and regulation for media organisations in this country	Confident	Not confident. **Revise** pages 124–125
13	Understand the relationship between celebrities and the media and identify a case study to research further	Confident	Not confident. **Revise** pages 134–135

Production

Production work is a core requirement for all Media Studies specifications. It is hard to study the media without getting the chance to create media texts and indeed that is usually what students find most interesting.

By learning how to construct a text for yourself you will gain a much better understanding of the production process and the ways it can be shaped by various factors such as budgetary restrictions and time limitations. By having to think about the way that your text communicates meaning to your target audience you gain a greater awareness of the way that other media producers have produced their text to have meaning for an audience.

You will probably only be able to choose from one or two media to produce your project and you may have to do so within a particular topic area or to a particular brief, for example. Therefore you will need to discuss this carefully with your teacher to make sure that you are doing the right project.

We have given you some general advice here about how to work within the different media and respond to the brief given by your teacher. Each school and college has different technology and it would be impossible for us to give you enough information about how to work with your particular technologies, so you will need to ask your teacher for help.

Remember that the higher your production values, the better your project will be. If, for example, you rush something together very quickly without thinking about your target audience or without thinking about the ideologies operating or what representations to use it will not be as successful!

Topic checklist

○ AS ● A2	OCR	AQA	WJEC
Practical production work	○ ●	○	○
Moving image production 1	○ ●	○	○
Moving image production 2	○ ●	○	○
New media 1	○ ●	○	○
New media 2	○ ●	○	○
Radio production	○ ●	○	○
Print	○ ●	○	○
Writing up your work	○ ●	○	○

Practical production work

All Media Studies AS and A2 courses ask you to produce practical work – after all, it is just as important to be able to create media texts as to analyse how others have created them and to understand the theories and debates relating to the production of these texts. It is also very important to be able to demonstrate that you have the skills to be able to produce texts yourself!

All this practical work must be accompanied by the required written work – which you can read about in more detail at the end of this section. Remember that you should be aiming to produce the best work possible – you may not have access to professional equipment or software but you can take time and effort to produce your work well. Media is all about presentation of ideas, as you know by now, so make sure you present your ideas as well as you can!

General preparation

Pre-production means 'draft' and 'planning'. So a storyboard for an advert or a script for a dramatic sequence are appropriate for pre-production work where that is asked for.

However, where the requirement is for production work, it MUST be fully realised. You cannot submit a trailer which is only half-finished, for example, and expect to get a very good mark for it! The first rule of production work is: **GET IT FINISHED**.

The production process

Media texts, whether they are being produced by Time Warner or for a practical project, go through the same stages before they reach an audience. This is what we call the **production process**. This starts with an idea for a text, created for a particular audience.

Every media text begins with an **idea**. You may start by deciding which medium you wish to work in, you may be given a specified brief, but you still need to have an idea. Once you have a clear idea of what the media text will consist of, you need to be sure who the **target audience** will be. Unless you have a specific audience in mind while you are constructing the text, it is unlikely to be successful – a TV drama produced for a target audience of 5-year olds is likely to be very different from the same story presented for an audience aged 50+, for example!

There are usually three stages to the production process, as outlined below.

Watch out!

Not all syllabuses require you to engage in the same kind of practical work. Sometimes you might be given a specific task to choose, sometimes you might need to relate your practical work to your theoretical work and sometimes you might be given a free choice of what to do. It is important that you check with your teacher about exactly what you are required to do.

Action point

If you have any questions at all about what you should be doing for your project, the person to ask is your teacher. They will have much more idea about what you should be doing and how you should be presenting it. They are also going to be the best person to advise you about what to do. After all, they are likely to know what equipment you have available at your centre and also where your strengths are.

Don't forget

Unless you have no choice, it is always a good idea to make sure that you have done at least one practice project before submitting your work for assessment. Many centres allow students to complete three practical projects over the course and choose the best one for submission.

Checkpoint 1

What is the USP (Unique Selling Point) of your product which would make an audience want to watch/read/listen to it?

1. Pre-production

This is a vital part of the process, and often the longest and hardest. Some of the things you need to do are:

→ Research the target audience, and match the idea to this audience.
→ Deconstruct existing media texts that are similar in form or content.
→ Plan each stage of production, with equipment, locations and personnel booked in advance.
→ Additional material (e.g. graphics/titles for the beginning of a video or music to use as a soundtrack) should be created/found at this stage, so that you have a clear idea of the identity of the product you are creating.
→ Create the storyboard, flatplan, layout and site map or other planning documentation so you know exactly what you need for your production. This may change but you need a plan.

2. Production

This is often the most straightforward part of production, and involves creating the raw data (moving image, text, images, sound, etc) that will constitute the final product. The key here is to be organised so that you know what you need to produce when, and to make sure that you keep to the schedule.

3. Post-production

This mainly consists of editing, constructing the raw data into the form that the audience will receive. It is during this part of the process that special effects and soundtracks are added to films, where a radio programme is multitracked, and where print pages are proofread and printed. If you did your planning carefully and got all the material you need, this is the most satisfying stage, as you see your idea come together. It is important to remember the target audience at this stage and make sure the text will appeal to them.

Examiner's secrets

Once an audience knows about the media text, it needs to get hold of it. It is important in your writing about the project to write about where and how you would distribute the text and why this would be the best way to reach your specified target audience.

Checkpoint 2

Have you got copies of all your pre-production planning that you can submit for assessment with your work?

Practice

One reason for completing at least one piece of practical work before submitting your project is learning to actually complete a piece of practical work. Especially if you have not done GCSE media studies, you need to have at least one run through – ideally in the same medium as you will be using. You will find that simply getting all the pre-production work done properly so you really know what you are trying to produce, completing the practical work and then learning how to reflect on your work takes practice. After all, you would not expect to turn up for one of the written papers without ever having tried to write an answer to one of the questions, so why expect to produce a piece of high quality production work, never having had a trial run?

Moving image production 1

Whatever type of moving image production you choose to create, the pre-production and production processes are very similar. If you are producing a film production project or a television production project, the differences come in approach, context, rationale and, to some extent, form. The process – as far as you are concerned – is the same.

Pre-production ●●●

Planning

The first stage in your planning is to decide on your idea. You are likely to have already made your decision about the form of your project, e.g. making an opening sequence for a film, a trailer, a television advert or the title sequence for a new crime series. You now need to:

→ Identify your target audience and brainstorm how to appeal to them.
→ Identify your genre and the genre conventions which you might use.
→ Decide on where your project might be distributed and when it might be scheduled.
→ Complete a location recce and a risk assessment for each location which you think you might use in the production, even if you do not use them in the end.

Storyboards

A well-designed storyboard will make your production much stronger. It means you can think about the plot, characters, genre conventions, camera angles, framing, mise-en-scène, editing and overall structure of your production before you begin. If you are filming a horror trailer, for example, and you have planned the woodland chase sequence and thought about the mise-en-scène of the different shots, the angles you want, the framing, use of genre conventions and the use of light before you go to film, the filming will be far quicker and more efficient. Editing will also be easier if you know that you have the right shots. Of course, there is nothing to stop you reworking the storyboard on location on the day but having a storyboard before you start will give you a much better chance of producing a good production.

The storyboard should:

→ show each shot in the scene in order
→ give some idea about framing and camera angle
→ give shot lengths (usually between 2 and 5 seconds)
→ show how the shots will link, e.g. with cuts or transitions
→ give information about the audio required for each shot (diegetic and non-diegetic).

You don't need to be an artist to do this, but it's no good drawing a whole stick man, if the shot is supposed to be an ECU.

Shooting script

Filming is not usually done in chronological order but for each location in turn. If you are shooting the opening sequence for a new crime serial for example, you would shoot all the shots in the police station together and all the car chase shots together, even though the final edit might switch between them. Not only does this save a lot of time in terms of travelling between locations and setting up, it also helps to make sure that the lighting and weather are consistent for each location.

Production ●●●

→ Always use a tripod unless you actually want a wobbly shot (perhaps a POV shot) and if you are after a smooth panning shot, for example, use a track or dolly system. It is easy to rig up a tracking system with a bit of ingenuity if you do not have access to proper equipment.

→ Professionals very rarely zoom in – use a sharp cut to an ECU to establish a reaction shot, say, or pan across a scene and then use a CU of something of interest.

→ Keep the shots short and use variety of length, framing and angle to maintain audience interest. Think about the genre conventions to structure this. Also bear in mind whether you are shooting for film or television. Television tends to use more close-ups and fewer panning shots (think about the fact that you are shooting for a much smaller screen).

Think about what is in the frame – *Blade Runner*, for example, relies very heavily on use of light and dark to communicate emotion and atmosphere, with shadows across faces and scenes shot in low light conditions to establish mood and danger. Equally, you can use light and dark to indicate characters' feelings or genre characteristics – a wood is far more frightening in the dark than on a bright sunny day. Use lights, or even filters, to create effects and atmosphere and don't forget to use unusual camera angles if you need them, to create particular atmospheres. A bit of imagination during filming can be very effective.

Example

Many directors like to design a complete storyboard for a production before they even start. Ridley Scott, for example, drew the entire storyboard for *Gladiator* before even beginning to plan filming.

You can even buy some of these storyboards or find them on the Internet! It's worth looking at some of them to see how much detail the directors have planned before getting anyone else involved in their production.

The jargon

Shot sizes range from *ECU* (extreme close-up) through *mid shots* to *extreme long shots*. An extreme long shot is common at the start of a film, for example, to set the context for the action. An ECU might be used to show a character's reaction to something said by another character in a soap opera. A common sequence of shots for television is shot–reverse shot–shot to sequence a conversation between two characters.

Don't forget

If you are only able to use one camera, it is worth experimenting with shooting scenes from different angles, if the light conditions will let you and the action can be consistent, so that you have different choices during editing. Professionals use lots of cameras for just this reason!

The jargon

It is very easy to create a different atmosphere for your production using *filters*.

If you are using lights for your scene, you may well have access to proper *gels* (sheets of coloured acetate) which can be attached to the lights to create a particular atmosphere.

Practice

While you are getting used to using the cameras and creating meaning through visual information, you might try creating a crash edit narrative. A crash edit is a name for a type of filming where all the shots are filmed in sequence and there is very little editing to be done. If you are working 'in camera' you might not be able to get your edit points just right but you can still create an effective narrative. Why not try creating a narrative based around a title like – 'A Stranger in Town' or 'He must believe me!'? There's only one catch – to make you really think about visual narrative, you can't use any dialogue in your narrative!

Moving image production 2

Check the net

Your teacher will be able to give you much more help with ways of editing. It is a skilled craft but the basic rules are fairly easy to grasp. There are also plenty of good books and Internet sites which can help you – a good site to start with is Cyber Film School at www.cyberfilmschool.com or http://www.dvmoviemaking.com for example

Action point

You can also learn a lot about editing by watching different films and television programmes and seeing how they edit the shots together. For example, the editing of a soap opera might be very different from the editing of a news programme. As you watch different programmes in detail like this and think about the editing being used so you will also gain in skills and knowledge which you can use for textual analysis exercises as well. This is a good example of how the different units of a Media Studies course will integrate because you use similar skills for all the units.

The jargon

Continuity editing is the name for the type of editing which we are most familiar with. It creates a continuity of narrative, without an audience having to watch events in real time. If you see a shot of someone getting into a car and driving off, followed by a shot of that person arriving somewhere, you will assume that they have driven to that place without needing to see the whole journey! A common pattern is to see someone waking up in the morning and then getting to work – missing out all the messy bits in between!

When filming, remember that there is a line in any given shot that splits it into two halves. If you are shooting a sequence that switches back and forth between two characters, for example, all shots must be made from the same side of the line, otherwise the scene will look unbalanced. You must keep the same side of the line as in your establishing (or first) shot. If you suddenly cross the line, the character who was facing right will be facing left from the new perspective, and since the other character is still facing left, you have two characters facing left, and if they try to talk to each other that could become very confusing for an audience.

Mise-en-scène

This simply means 'what you can see in the frame' and is important for creating atmosphere and setting the scene. It's no good trying to suggest that your scene is set in a school at night if there are clearly lots of people moving around in the background between lessons. If you are trying to suggest that a location is very run-down, think about using a few props to make it look run-down.

What clues suggest the mise-en-scène here?

Editing

→ A **cut** is most common type of transition (a way of linking two shots in a sequence) used in film and television work between shots.
→ A **fade** can be useful to suggest the passing of time or a change of place/action.
→ A **dissolve** is occasionally used – at the beginning or end of a dream sequence, for example.
→ **Wipes** and other transitions available on most non-linear editing systems are very rarely used, except perhaps when creating an alien effect or to suggest a character descending into a nightmare, for example. (The exception is sometimes a pop video, where a lot of effects can be used very creatively.)

When shots are being linked together there are various ways of suggesting continuous action. It would be very boring to film everything that happens, so continuity editing allows the brain to fill in the missing bits in the sequence. Editors will often try to **match** shots, so, for example, they will match on a movement (e.g. two different characters walking in the same direction) or on an action (e.g. two different characters having a phone conversation).

→ A good way of keeping the action continuous can be to use a **cutaway**, where you keep the soundtrack running underneath all the changes – a conversation continuing as the shot–reverse–shot happens or a voiceover continuing over a close-up of a key moment, for example.

→ Editing can also create **atmosphere**. If you were watching two people having a conversation, you would automatically switch your focus between them. Good editing mimics this but controls your perspective and hence your opinion about the action. Depending on the editing, the audience can feel sympathy for either person through use of camera angles, shot length and what reactions are shown to the audience, for example.

→ Editors also take care to control the **tempo** of a scene, building tension slowly at the beginning of a sequence, gradually making the shots shorter as the tension builds and using a very quick shot–reverse–shot sequence during an argument, for example.

Graphics and overlays

Your finished project is likely to need graphics and overlays. You should aim to keep all overlays simple – a simple banner at the bottom of the screen, giving the name and occupation of a person being interviewed in a documentary, is as much as you need. Complex *Eurotrash*-type overlays running across the bottom of the screen are usually unnecessary!

Documentation

You will have to produce written documentation to support your project. There is general advice about how to do that at the end of this practical section.

> **Don't forget**
>
> If you are using basic editing equipment, there may be no facilities for adding graphics to the final edit. There are various imaginative ways around this, such as writing/drawing the graphics on a whiteboard and then filming the whiteboard and treating the shot like any other shot.
>
> Some linear systems allow you to design titles or use overlays to some degree so you will need to experiment to see what can be done.

> **Practice**
>
> To make yourself familiar with editing techniques a good exercise is to video record the opening section of two relevant texts, e.g. two horror films or two sitcoms. Then, in your group, sit and make a list of each shot used in this sequence. (Don't study more than 2 minutes of it or it will take you forever!) Look in detail at camera angle, shot length, mise-en-scène and editing between shots and ask yourself what meaning is being constructed in each shot. Not only will you soon begin to pick out typical conventions, this type of close reading will also really help with your textual analysis work. And as you begin to look more carefully, you will begin to see what conventions are used in different programmes. *Eastenders* is very fond of using matches on action for example . . .

New media 1

Check the net

The best known place on the Internet to find out about less successful webpages is www.websitesthatsuck.com
There are many good sites to compare – you might start by looking at www.bbc.co.uk or even www.lastminute.com, for example.

This section focuses on new media projects, such as websites, computer animations, simple computer games or multimedia CD-ROMs. There are many exciting projects which you can produce here and one big advantage of creating such projects is that you may have access to the necessary software at home (or be able to write your own code), so you may be able to work away from school or college more easily than with other media. However, if you are not already familiar with this type of software it can be quite complicated, so you are advised to discuss your project carefully with your teacher or lecturer before you begin.

General planning

Whatever the project you are creating, you need, as ever, to start by having an idea and then thinking about your target audience.

Identify your target audience

Identify the target audience for your site quickly as this will dictate what you create and how you design it. A site aimed at 4–6-year olds will be designed very differently from one targeting 45–50-year olds. A site designed to sell up-market holidays will be very different from a webzine presence set up by a magazine such as *Loaded*, for example.

Once you have started to plan your site and defined your target audience you should carry out some research, to identify how best to appeal to this audience on the web by:

→ identifying sites also targeting them
→ looking at how they have been designed and laid out.

This research can prove an important part of your evaluation.

Appeal to your audience

Your site must be interesting and easy to understand. It is no good if users cannot work out what the site is about or which links they need to click to get somewhere. You should also make sure that your graphics are appropriate and fit your site. In other words, you need to make sure that it is communicating with the reader. Just as a magazine front cover or the title sequence of a TV programme sets the tone of the programme to follow, so a site's opening page should establish its position quickly – background colour, layout, use of images, etc, should all work together to create the right identity for your site.

If you look at some of the best sites, you will notice that there is a clear sense of branding on all pages; navigation is transparent (that is, easy to grasp); the pages are uncluttered and straightforward and they load quickly. It is often said that websites benefit from having lots of pages with little information on them rather than a few pages containing lots of information!

Examiner's secrets

Identifying your target audience for your project very clearly will really help you create a strong product. It is important to very explicit about this audience in your writing.

Basic planning

You should always start by planning out your site on paper so that you know how many pages you will need and which pages will link with each other. Once you have planned the basic outline, it can be useful to create a 'navigation pane', which is a list of links in a table which you can put on the left-hand side of the page and copy onto each page. There are many different ways of doing this, but if you can put this element on each page, it will save you a lot of time while making the pages.

Design issues

To design a really effective website, you need to start by imagining how the audience will react to the site and how they will use it.

→ Will they expect the links to be obvious?
→ Will they want lots of graphics or will they want more information on each page?
→ Make sure that there is a home page link on every page of the site.
→ Make sure the user can email someone if they have any questions.
→ Think about what the user is likely to want from the site.
→ Are they expecting to be entertained?
→ Do they want to find out a lot of information?
→ Is this site like an online magazine?

Most importantly, as with other media products, you should start by thinking about what your USP is. That will dictate the form of the site and the layout of the pages. A site giving information about club nights in a local area is going to have a different USP from a site giving information about which clothing manufacturers have links to Third World sweatshops. Therefore the content, layout and navigation are going to be very different.

Layout and design

Try to use only two main fonts on any page (just as with print products). Sans-serif fonts are better than serif fonts as they are easier to read on a screen. You should also think carefully about the size of the font and also the colour scheme. Pink text on a sky blue background can be quite hard to read!

Do not try to be too clever with your pages – if you are not used to creating complicated effects such as drop-down menus, it is probably best not to use them. Simple pages which work and which appeal to the particular target audience will invariably be more successful than pages which do not always work.

Don't forget

You do not need to produce live pages for any of the exam boards at present because it is impossible to mark fairly from live sites.

Action point

If you really want a particular font for a heading, for example, create it using a graphics package (such as Photoshop) and save it as a JPEG file. Then, all you need to do is put it into the page as a picture and the font will work. This is better than using an unusual font in your webpage as it means that the font will work properly on all computers.

Practice

You should take time to look at existing webpages or existing games created for a particular audience or a particular context. Try to look at three similar products – what common codes and conventions can you identify? What USP does each product have? What is the relationship between text and audience in each case?

New media 2

Don't forget

Many image editing programmes, such as Photoshop, have a 'Save for web' option which will show you how long it will take for your images to download and allow you to change settings to make the images download more quickly.

Watch out!

When you copy your work onto a CD-ROM, make sure that you have put all your pages, images and other source materials inside a single folder before you copy them. This way, you should find that the pages will display the images correctly. If they do not, you may need to ask someone to help you edit the code to make sure that the images will display properly.

Some basic design advice for webpages

→ Backgrounds should be simple and usually a single colour.

→ A background image usually makes it quite hard to see the writing on the pages.

→ Keep a pale background so that it does not 'jump out' at your reader. Text is usually best laid out as if for a print document (although remember the sans-serif font) and it is best not to use too many sizes or effects.

→ Links should be obvious (usually they are underlined). There is not a lot of point in making a link but not showing the reader that it is a link!

→ Check the size of images for your pages. It is very easy to create very big files if you are uploading pictures from a digital camera, for example, and if you simply put these onto your page, they may take a very long time to download or even crash the computer.

→ If you are submitting your work on a CD-ROM, make sure you put copies of the images onto the CD-ROM as well or they will not show when someone opens the site. Make sure they work before submitting your project!

→ Webpages should scroll vertically (up and down the page) if necessary but never horizontally (across the page) as this makes it very hard for your reader to follow the pages.

→ Always design for a fairly small monitor – for example, a page which is 760×420 pixels should work in a browser window on a normal 15″ (375mm) monitor without problems.

→ It can be helpful to use a table and put text into different columns on a webpage – it can make it easier to read the text. You should also make sure that, if possible, the text is broken up with images as it helps the reader.

→ Buttons are useful to make interesting navigation but you should always use text navigation as well for accessibility.

→ Photos should normally be saved as JPEGs and you should keep the file size down as much as possible. If you have drawn a cartoon for your site, made a diagram or drawn a heading, for example, it may be better to save these images as GIFs as they usually have a much smaller file size.

Basic computer games

Most of the advice given above about websites is directly relevant if you are creating a new computer game. You are not going to have the resources to produce a fully realised *Grand Theft Auto*, for example, but you can choose to create a simple game or to create the first level of a more complex game.

Once you have decided on your idea and identified your target audience you can begin to plan your game. Again, the process is the same for a simple game or for the opening of a more complicated game. You will need to decide what the game will look like (the environment), the events (the game progression), the characters or objects which the game player will interact with and how the player will navigate and

interact with the game. Again, you are unlikely to be in a position to develop a complex game with extensive interaction and choices of strategy, but you can produce a basic scenario with choices and interactivity to demonstrate an understanding of the process.

Flash animations

One programme which is often used to create some excellent games is Macromedia Flash. This is the programme which creates some of the most effective animated webpages and 'splash' pages you will see on the Internet and is also used to produce animations and games. If you want to create a game for inclusion in a webpage or to promote a product or a new film or even to teach something to someone, Flash is often the best tool, as it creates simple games fairly quickly. This type of game would be a good choice for a media project and you may well be able to create a game for a new film, for example, and embed it into a website to promote the film on one of the main pages.

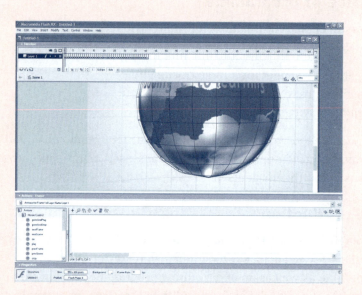

The Flash interface

Watch out!

Each of the exam boards has different requirements for how you should submit your work. You should check with your teacher what format your work needs to be in for moderation.

Example

Of course, as the technology continues to develop, there will be more opportunites to create these sorts of projects – what about a game to promote a new brand of trainers which can be played on a 3G phone with your friends, for example? However, much of the technology is complex and there can be a lot of problems with getting it to work (as you have probably found when trying to get to websites or play games). So, unless you have been taught what to do or have already taught yourself how to use these technologies, it is probably best not to use them for your project.

Practice

If you decide that you are going to create an animation or a computer game for your practical project, you should start by thinking very carefully about who the target audience will be, what your institutional context will be and therefore what media language and representations you will employ in your product. You might, for example, decide to create a simple demo game for a new release of an Electric Arts game, such as 'SimSchool'. In this case, you would need to research the rest of the *SimCity* games and think carefully about how you would design 'SimSchool' to fit this institutional context and appeal to an appropriate target audience.

Radio production

Radio can be a very successful and interesting option for practical work at this level. However, you should make sure that you take the time to research radio production and the radio industry before you start your production work. We tend to listen to a lot of radio as 'background' to other activities and therefore we have a lot of knowledge of the forms and conventions of radio work but we tend to take this for granted. If you are creating a radio programme for children for a local radio station, you will construct it very differently from Radio 4's *Today* programme, for example. Moderators often complain that students do not always identify, analyse and use the appropriate codes for their institutional context and audience when creating radio work.

Choice of programme

Successful choices at this level for radio production work might include:

→ creating a segment for a radio documentary
→ a section from a local radio breakfast show
→ a radio play based on a short story.

If you decide that you want to create an extract from a 'music radio' programme, remember that you will be assessed on the links around the music, not the music itself.

Pre-production

To do a radio production, you probably need access to a good-quality recording device such as a good minidisk recorder and a good microphone as well as an audio editing programme to allow you to manipulate the material and use additional tracks – for example, to have music underneath some of the words. If you can get access to a computer with audio software such as Cool Edit Pro, you will find it easy to multitrack.

Planning

Each genre of radio production uses very different codes and conventions. For instance, local radio might also use sponsorship and adverts during programmes. You need to identify your chosen genre and station early on in the planning process, to make sure that you have time to research these codes and conventions to use in your own production work.

Following on from this, you should research your target audience for this programme. Remember that for radio, target audiences are usually very different and often quite specialised:

Don't forget

If you are creating a music-based programme, why not simply fade the music tracks in and out so you can concentrate on the linking sections which you will be assessed on – then your moderator does not need to listen to all the music as well, making it quicker to asses your work!

Watch out!

Not all the exam boards allow you to choose your own production context – you need to check with your teacher about what sort of radio programme to make.

Action point

It is possible to make a radio production simply by using two cassette players, providing you plan and rehearse carefully before you start recording. If you can use the volume control as a fader you will help to avoid clicks as you record from machine to machine.

The jargon

To *multitrack* your programme simply means to have more than one track playing at a time. So, for example, the music which leads in to your narrator speaking might overlap for a few seconds and there might be sound effects being used underneath the narration to provide atmosphere.

→ Local stations may well have different audiences at different times of day.

→ Formats will vary at different times of day.

Content

Once you have identified your institutional context, the target audience and the genre of programme, you can begin to think about the content. Some bits of your programme can be recorded individually, such as jingles, and some sections can be edited together before being put into a programme, such as a series of interviews on the same subject. The final narration or content is probably best scripted to give your presenters some support, but try not to let it sound as if it is scripted.

As part of your research, you should have noticed how varied radio tends to be, with:

→ frequent changes of voice or content
→ short segments
→ lots of jingles/adverts/recaps, etc
→ frequent references to 'later in the programme'.

which all help to make the listener feel secure and engaged. Your material needs to work the same way. If you are producing a segment with a single voice, for example, use sound effects below the voice to create a sound picture. Remember that your listener has to be able to focus on the material, so keep things varied for them.

You also need to think about how you can clearly establish the USP of your programme. If you ever try listening to radio stations from around the world, you will realise that even when you cannot understand the language being used, it is easy to identify the genre and USP of a radio programme very quickly. This is a sign of a well-made programme so try to make your programme distinctive and the genre apparent!

> **Don't forget**
>
> Exam boards are usually pretty understanding if you not been able to use high-quality microphones for recording interviews, for example, but will be less sympathetic about poor sound quality which could have been avoided, such as noises from the interviewer getting in the way or rustling noises because someone was fiddling with the mike during recording.

> **Practice**
>
> One way of getting used to using radio equipment can be by recording a simple five-minute programme such as 'A day in my life'. There are several key elements to this as an exercise which will teach you a lot about the equipment. Firstly, you will have to make decisions about what to record, if you are to get a whole day into just five minutes. You need to learn to use the audio equipment, recording and uploading the material and editing your material. You will also learn a lot about recording material for use in a radio production and about how different conditions affect the quality of your recording. To help you get used to multitracking you can think about what music and sound effects you will use to support your programme and experiment with the effect the different tracks can create. You could then try doing 'A day in the life of the school/college' so that you also get used to doing interviews, recording on location and using a different range of voices and content.

Print

Print production work remains a very popular option for practical work – especially at AS level. However, it is not a good option unless you have access to some basic facilities:

→ A computer with some DTP software (try not to use Word if you can avoid it, as it makes layout awkward).
→ A printer which can print in colour (unless you are producing black and white texts such as a newspaper).
→ A camera to take some original images to use in your text (ideally, a digital camera, as this makes it easier to get the pictures into your pages) or some drawing/illustration software if you are going to design your own images.

Planning your artefact

It is very important to define your target audience well before you start to construct your project because the target audience will affect every design decision you make. For instance, the target audience for a new newspaper will significantly influence the font chosen for the masthead and the audience for a promotional campaign will make a big difference to the representations used in the posters.

Your next job should be to produce the flatplan in line with your decisions and the overall look of the product. Don't forget to think carefully about where these products will be distributed – the shape of a billboard poster, for example, is very different from that of a magazine advert.

You also need to do some research into similar products.

At this point you should be able to list the images which you are going to produce, the copy you will need to write and the other elements you will need to design for your project. Once you have planned when you will do each of these things and checked that content and form still match target audience and suggested institutional context, you are ready to start the production phase.

Submitting your work

It is always worth testing your product on the target audience during and after the production process. Keep a record of their comments and use them during production to improve your product or afterwards, in your evaluation, when writing about the success of your product.

When you produce your work for moderation, make sure that the names of all members of your group appear on all the work, along with the centre name and number. When you print the work, make sure you use a good quality printer and good quality paper. If you are producing any kind of glossy publication, it makes sense to use glossy paper. If you have created billboard posters, make sure you print them landscape. It does not matter if you cannot print up to the very edges of the paper with your printer, even if your publication should not have margins, but it does matter if a broadsheet newspaper is printed on a piece of A4 paper.

Design tips

→ Keep your pages as uncluttered as you can and don't run text right up to the margins. Remember that a picture speaks a thousand words and that long sections of text can be broken up in many ways. Always relate any design decisions you make to the target audience and to codes and conventions used in similar publications – unless you have made a deliberate decision to be different!

→ When taking photographs think carefully about the mise-en-scène as well as the camera angle, framing and lighting to create the effect you want. It's much easier to get it right when you take the photo, rather than have to try to manipulate it extensively later.

→ As with the images, be careful with your copy. Make sure the vocabulary, sentence structure and content match with your target audience and institutional context. An article about a film star in *The Times* is going to be written very differently from an article in *Loaded*. If you are creating a poster to advertise a new action/adventure movie, you are unlikely to use images of flowers.

→ Remember that your purpose is to provoke some kind of emotion in your audience. You are not producing factual texts but texts which affect a mass audience. An advertising campaign must motivate the audience to buy the product, support the charity, etc. A newspaper does not seek simply to report the news but to make judgements on it and to make the reader react to what is reported – with anger at a dreadful crime or with sympathy for a celebrity having difficulties in their personal life.

The jargon

A *flatplan* is a rough layout for a magazine or newspaper. It is simply a series of boxes, each representing a page in the publication. Each page is mocked up in terms of where the images and copy will go, to get an idea of how the final pages will look. This also enables the editor to place advertising, editorials or other content, to decide how many images will need to accompany each article and to get an idea about how much space each article will need.

The jargon

House style – Where publishing institutions produce more than one title, they often use the same house style across a group of titles so that the audience can easily recognise them. For example, you will probably have noticed by now that all these Pearson revision guides are produced to the same house style, regardless of the subject.

Watch out!

If you are producing a product which is bigger than A4, you must be careful to work at the right size. If you produce an A4 size broadsheet, the pages would be distorted if blown up to the right size, and therefore not fit for purpose. If your poster is intended for a roadside billboard, it should not contain extensive text as the target audience would not be able to read it. Again, this would not be fit for purpose!

Practice

It is always worth experimenting with the equipment and software before beginning on your final project. After all, you can't concentrate on designing a high-impact advertising campaign if you are still trying to work out how to get the images from the camera onto the poster! For example, you could take pictures of four of your friends or family in various poses, with different lighting conditions, and use them to create a poster advertising a new film in a particular genre, e.g. a horror or a romantic comedy.

You may find that you can use the same photographs for different posters by manipulating them to change the lighting, add effects or crop them differently. This will help you think about genre conventions, institutional codes and conventions and media language.

Writing up your work

Examiner's secrets

If you can include examples of similar texts which you have studied as part of your research, this can be very useful. You can include these in an appendix perhaps – maybe even marked up to show that you have deconstructed them!

Watch out!

You may be required to include evidence of planning, such as storyboards or flatplans. For all the boards, you will be expected to provide evidence of research into similar texts. Again, you need to check with your teacher about the exact requirements for your unit.

Watch out!

Usually there is a specific requirement to demonstrate some level of theoretical study to support your practical work. Make sure that you can write about genre, institutions, products you have studied and codes and conventions, for example, to demonstrate how much work you have done!

Although the requirements for the format and length of the written work to support your practical work vary between the boards, the basic principles and content are the same. The object of the written work is to demonstrate the levels of understanding, research and engagement with the project. A well-written evaluative piece will reflect the level of knowledge, understanding, research, preparation and commitment which have gone into the project. It is a chance for you to show how much you understand about the medium, the genre and the relationship between text and audience for your chosen brief.

It is not necessary to keep a production diary for your project. A production log may be required as part of the written work, but even the word 'log' should not be taken as meaning 'diary'. It is all too easy to give a detailed account of which bits you edited on the last Monday and the problems you had because the computer kept crashing, but you are not going to get marks for this. Writing about the problems that you had in choosing the very best shots to make the target audience identify with the main character immediately and why you made the final choice that you did will get you lots of marks. It's to do with understanding – the first account simply describes what you did but the second version explains what you were doing, why you were doing it, what you were trying to achieve, why this was important and how you overcame problems. Which way do you think you can show your teacher or a moderator how well you understand what you have been doing?

Writing up your work ●●●

Pre-production

You will be required to write up how you researched and prepared for your production. Generally, your research should show all the preparation you have done for your project. If you don't include a storyboard, how will anyone know that you completed one? If you took time to study a range of Westerns before constructing your own opening sequence, how will your teacher (or a moderator) know how much you learnt about genre conventions and the ways that Westerns usually start?

A vital aspect to your research and planning is, of course, your target audience. You need to demonstrate that you are aware of what the target audience for your project might be, how it affects construction decisions and in what ways you have made your text directly relevant to this audience. After all, this is a fundamental aspect of research and planning, since you need to have made all these decisions before starting on construction, so why not take the time to demonstrate this in your writing?

Production

You need to write about the production phase as well. One of the most common mistakes people make when writing this section of their report or evaluation is that they turn it into a diary.

A 'process' approach ('what we did today') does not demonstrate your understanding of genre, media language, representation, audience or institution. Always try to make sure each point you make about the production process demonstrates the application of one of the key concepts to your product.

Reflective commentary

Again, this may be called different things by the different boards but you will be expected to reflect upon the product you have made and compare it to existing products. You should try to write about:

→ how you have used typical codes and conventions from your chosen genre
→ how your product relates to existing commercial products
→ how you have made sure your product is appropriate for its institutional context
→ how and where your product would be distributed/shown and why
→ key aspects of the media language which you have used
→ the representations you have used
→ the ideology of your product in relation to cultural and institutional context
→ the relationship between your product and your target audience.

It's a good idea to test your product on at least a couple of members of the target audience – what responses and comments did they have about the product? What strengths and weaknesses do you think your product has?

And lastly, we often encourage students to finish by writing about what they would do differently if they were to do the project again. Here, as with earlier comments, you have to remember to keep a media focus to what you say. You are not likely to impress your teacher or a moderator if your reflection finishes with the statement:

If we were to do this project again, I think it would be very helpful to learn how to use the equipment first because we did not really know what we were doing most of the time.

Don't forget

Remember that it is not necessary to submit a video tape for every member of your group if you have done a film or television project, for example. A single tape for the whole group is fine – but make sure that the FULL name of all your group members is on the tape. A moderator may not be able to work out who 'Kaz' and 'Jonny' are!

Examiner's secrets

If you are producing print artefacts, try to get them laminated if you can. If you can't get them laminated in school or college, try a print shop. This does make sure that your work does not get spoilt or damaged in any way during marking or moderating – if you put a sticky label on the back with the names of everyone in the group, it also helps.

Practice

Sometimes students hand in written work like this:

On Monday we could not get the computers to work properly so we were not able to edit the scene properly. We spent a lot of time talking about what music we might use as a soundtrack but Kerry had forgotten the CDs so we couldn't do anything. When she did bring it in, it was no good because the track finished at the wrong time, before the credits finished showing.

Of course what this student COULD have written was:

We spent a great deal of time thinking about which track on the ambient jazz CD we might use as a soundtrack. This is appropriate music which is typical for the opening sequence of a programme in this genre, such as in XXXX. However, we had to edit the track to be sure that the climax in the music matched the climax in the action.

Now reread what you have written and make sure it is more like the second example!

Planning table

Medium	Example text	Planning may include	Production work	Evaluation suggestions
Film	Film trailer	Audience research Storyboard Genre analysis Production company plans	Shooting material Editing for trailer – teaser/video/general release? Mise-en-scène, framing, camera angles, etc on storyboard for production Soundtrack devised and added	Evidence of study of similar trailers Justification for construction of trailer Identification of key signifiers/contextualisers for audience/values/representations/institutions
	Opening sequence	Audience research and genre research Planning for film and form and function of opening sequence Graphics, titles and other overlays or elements to be integrated into the opening sequence Selection and recording of soundtrack to 'back' action, as required	Filming of action – both for pre-title sequence if being used and main opening sequence Filming of action to show under graphics, etc, for main title sequence, as required Editing to construct meaning, establish genre and hook audience Integrating graphics, action and soundtrack, as required	Evaluation of how successful sequence is as an opening sequence Identification of key signifiers for audience and genre expectations Analysis of how sequence establishes narrative and context Comparison with equivalent commercial products Testing on sample audience
	Short film	Audience and institutional research Storyboard Scripting/shooting scheduling	Shooting required scenes Creation of appropriate ideology/theme Editing to construct Realising as complete artefact with titles and credits	Study of short films and justification for chosen genre/approach Evidence of detailed study of film craft
Television	Title sequence for new programme	Market research Storyboard Genre/character/scheduling and content of new programme identified	Construction of images Construction of representations, e.g. characters Construction of graphics, e.g. overlays Construction of title sequence Choice/creation of audio/soundtrack	Defining context of artefact to reflect genre/programme identity/key themes/characters and linking visual/audio Evidence of study of equivalent programmes Research regarding chosen station and scheduling
	Extract/package for news programme	Storyboards Scheduling/channel and audience research	Use of appropriate forms and conventions to frame action Links in and out of material Shooting of sequence Editing of sequence to fit exact context and placement	News values deconstructed and Regional/local/national characteristics – examples Target audience identified and product related to them Evidence of detailed analysis of current news production
	Opening sequence for new children's magazine programme	Storyboards Creating an identity for the programme and constructing appropriate presenters Creation/selection of title music/graphic elements, etc Choosing and recording of suitable soundtrack, if required	Filming of studio sequences Filming of reported sequences/outside broadcasts, etc, as required Filming of sections using studio audience, if required Editing to create meaning Integration of soundtrack/graphics and other elements to establish clear identity and content for programme	Research into children's factual broadcasting form and function Analysis of equivalent programme in terms of content, presenters, style and identity Identification of target audience relationship to content Testing of final product on sample audience
	Opening section of new documentary	Storyboards Initial research into suitable content Consideration of documentary issues, e.g. form, use of presenter, content, etc Construction of graphic elements, e.g. titles or overlays Choosing of soundtrack, if required	Filming of content, e.g. to camera or 'mantle of the expert' sequences Filming of content – studio-based or on location, as required Editing to construct meaning and establish documentary form Construction and use of additional graphics, titles, etc, to clarify and contextualise subject as necessary	Analysis of documentary form and function Explanation of suitability of chosen form and presenter/voiceovers as required Comparison with contemporary products Deconstruction of documentary and how it constructs meaning Evaluation of success of product – probably following testing on sample audience
Print	New tabloid paper	Market research – niche? Identification of suitable political and ideological position Careful identification of specific target audience	Flatplan completed and production scheduled Stories written Images constructed Page layout completed Artefact printed/copied full size	Analysis of competing titles Research on circulation Critical reading about news/journalism/newspapers Identification of target audience for new tabloid and position of newspaper Live testing on sample audience
	Comic for primary school audience	Market research – niche? Analysis of existing comics Primary research – gender-divided audiences Flatplan	Detailed flatplan Project planning and schedule Cells drawn Pages constructed Title page/other pages designed Pages collated and presented as an artefact	Research history of comics Analysis of audiences Identification of genre features for target audience Development of new cartoons and design of comic Deconstruction of similar products Popular culture debate regarding comics as print texts
	Series of print adverts for a charity	Analysis of existing campaigns in terms of methods and distribution Research into chosen charity and how to promote the charity Creating a theme to the campaign and a campaign identity Creating graphics, layouts of final adverts and planning shooting, etc	Taking photographs as required Construction of necessary elements, e.g. logo DTP of final adverts – copy and images (campaign to consist of 6 adverts) Printing of adverts to scale (NB portrait or landscape as appropriate)	Research into charity promotional campaign techniques Deconstruction of similar campaign Analysis of how material targets particular audience Consideration of how and where these adverts will be displayed Testing on sample audience
	Print advertising campaign for a relaunched product, e.g. chocolate or trainers	Analysis of existing campaigns in terms of methods and distribution Research into chosen product and how to promote the product Creating a new brand identity for the relaunch Creating graphics, layouts of final adverts and planning shooting, etc	Taking photographs as required Construction of necessary elements, e.g. logo DTP of final adverts – copy and images Brand identity evaluated and sustained throughout campaign (usually 6 adverts) Printing of adverts to scale (N.B. portrait or landscape as appropriate)	Research into advertising techniques Analysis of target audience characteristics Deconstruction of similar campaign Analysis of how material targets particular audience Consideration of how and where these adverts will be displayed Testing on sample audience

Medium	Example text	Planning may include	Production work	Evaluation suggestions
Radio	Alternative music programme	Audience research Scripting Musical genre identified Presenters researched Comparable stations researched	Tracks selected for inclusion Identification of presenter and aural clues Creation of jingles Additional elements, e.g. news flashes/adverts Station ident created and integrated	Evaluation of comparative stations and programmes Analysis of different audiences for different stations and different audiences Establishing audience loyalty Engaging with audience and linking music Maintaining artistic structure of programme
	Extract from radio play	Market research Scripting/sourcing play	Script Performance Sound effects Soundtrack/characters Recording of performance Using multitrack to construct, play and add elements such as titles and closing sequence	Evidence of planning for completed production Research into history and form of radio drama Research regarding chosen station and target audience Live testing on a sample audience
ICT/New media	Website for new film release	Market research Site map designed	Graphics, audio visual and animated elements created as required Basic layout template designed and elements integrated Sense of identity between film and promotional material established Website created and tested Website links checked and site uploaded for beta testing Design of key elements for film, e.g. logo, mock-ups of stills, cast list as required	Research about web-based film marketing Research into how to promote the site and link with film Awareness of issues relating to web-based production Comparison with existing products Live testing on sample audience
	First level of new computer game	Market research Game plan design Mocks for screens	Static and animated elements designed Game path created and optional paths tested Environment developed and game integrated into environment Supporting materials (e.g. help book or scenario leaflet) designed as required N.B. It is NOT expected that a game produced for an A level project would be fully realised but the game should be workable in some way – in a similar way to demo versions of games which are freely distributed as promotional devices. Title sequence or graphic sequence designed to start game	Extensive evaluation of contemporary products Clear identification of genre, narrative and target audience as well as technological issues Careful identification of game scenario, procedures and protocols, including clear mapping of levels and progression Deconstruction of characters constructed for the game or user-identifiable elements Live testing on sample audience
	Launch edition of new ezine	Market research Identification of appropriate content and layout Planning and layout for ezine – storyboard or flatplan Consideration of distribution issues and how they will affect quality/structure (e.g. bandwidth)	Assembly of content, e.g. images, copy, animations, graphics, photographs, etc Construction of ezine Design/structure to target audience	Research into new media ezines against traditional magazines and fan publications Detailed analysis of contemporary products Examination of ezine in terms of target audience, content and possible costs Testing of finished product on sample audience
Cross media	Music 'promo' video	Market research Identification of musical genre/track to use	Filming of live performance from a variety of angles (ideally mimed to a recorded version unless good quality audio recording equipment is available simultaneously) Recording studio version of track recorded first (this could be used for group to mime to during filming to ensure continuity, for example) Filming of narrative or 'creative' sequences to intercut with performance Filming of other elements as required Editing of performance, narrative and other elements for final video Graphics, overlays or titles added as required	Deconstruction of final product in terms of different elements and how they construct meaning Comparison with equivalent commercial products within chosen musical genre Research regarding genre elements and appropriate constructions, e.g. costume and performance style Consideration of how and where the video will be released and its primary purpose Research into form and function of music videos Testing on sample audience
	Animation	Market research Preparation of models/sets or designs for characters/elements to be used in computer-designed animation Storyboard Identification of target audience and context of distribution	Filming of animation (stop motion) or use of computer animation package to create artefact Editing to create meaning Construction of characters/elements to appeal to target audience and establish clear narrative and genre Title sequence and closing sequence completed	Research study of stop motion animation Detailed analysis of 2 animated films Evaluation of animation products and film Analysis of how product constructs meaning and how successfully genre/narrative conventions have been employed Testing on sample audience
	Multimedia promotional campaign, e.g. for new computer game or style of jeans	Market research – analysis of target market, appropriate promotional methods and form and function of appropriate materials Preparation of materials, e.g. photographs/filming of people wearing chosen jeans, title sequence/soundtrack/graphics for new game Brand identity created to distribute across these different media Research regarding ezine marketing, viral marketing, pop-ups and other web-based promotional devices	Design of product and badging for relaunch Design of elements for different multimedia contexts and identification of cross media elements Construction of adverts in each of the different media to be used Campaign identity established, appropriate for audience and branding	Research into new media advertising strategies Detailed analysis of contemporary products Examination of chosen promotional methods in terms of target audience, content and possible costs Testing of finished products on sample audience

Production Dos and Don'ts

These are probably some of the golden rules for production work – some of them relate to all media, some only to one medium but all of them are based on experience of things that can go wrong.

1. A planned production is always better than one you make up as you go along.
2. If you have no plan of which shots you want to film on any one occasion you will probably end up forgetting one and having to go back again.
3. If you are recording a radio interview, make sure the microphone is far enough away for it not to bang against your mouth when you are speaking.
4. Also always check your sound levels before you begin recording. For radio work or moving image work, poor soundtrack is one of the most common reasons why moderators criticise student work. It's not hard to plug an external microphone and headphones into a video camera, for example, and use them to record only the sound you want. Equally, if you are recording a radio interview you don't want to get home only to find that you did not have the sound levels turned up high enough.
5. If you are using software, such as video editing software or web design software, for a project make sure you save your work each time. We advise students to keep two running backups so they save over the top of the oldest of the two copies each time. That way you are not filling up hard drives with lots of copies of the same thing but you've got a spare copy in case something goes wrong. We also encourage them to burn a copy off to DVD or CD at least once during the production process – just in case. It's better to have a couple of CDs you don't need than to lose all your work – and it does happen!
6. If you are going out on location for any kind of source material, e.g. taking photographs, recording a radio interview or shooting footage for a film, always do a few basic checks before you go:
    →   Check that you have got all the leads you might need, at least one set of spare batteries and at least two tapes (with your names on). Then double check this. Ensure that someone has a detailed list of everything you are taking (down to the last cable) and make sure you bring it all back again.
    →   It's important that someone knows where you are going, what time you will be back and what you are doing. Also make sure that someone has a mobile phone with them and ideally leave the number with your teacher so that they can contact you if there is a problem. Never go to do an interview or do a shoot on your own – just in case.
7. Always leave enough time for post-production work – it usually takes at least as long as the production phase – and don't start that until you have finished pre-production!
8. Make sure that you have got far more material than you need and be prepared to be ruthless during editing. Better to not use lots of good stuff than have to pad out what you have got.
9. Allow enough time to write your work up well, including lots of reference to relevant contemporary commercial media texts to demonstrate how well your product matches with them. If you get the time to get some audience feedback as well that often helps! And remember not to write about the production process – teachers and moderators already know how the software works!

In this section we give you some advice about how to revise for Media Studies exam, write different types of answers, undertake research and use resources like the Internet to help you.

Media Studies is a very wide-ranging subject, requiring different skills from you, ranging from those of media producer through media analyst to critical theorist. It is interrelated with many other areas, such as sociology and psychology and study in this subject therefore usually centres around the use and understanding of the key concepts rather than any set requirements for different units. So here we have suggested ways that you may like to structure your revision, ways of preparing and writing critical essays and developing the skills you need to write about Media Studies. As frequently pointed out in this book, however, Media Studies is also such a broad subject that you need to be doing a lot of this all the time – not just when you are making final preparations for an exam. Critical autonomy comes from a wide experience of many different media texts and you can only gain that by making the effort to read, watch, play and engage with a wide range of media texts every week and being prepared to talk about them in class to test your opinions and practise maintaining a debate.

Topic checklist

○ AS ● A2	OCR	AQA	WJEC
Revision techniques	○●	○●	○●
Critical essays	●	●	●
Exam questions	○●	○●	○●
Independent research	●	●	●
Using the Internet for research	○●	○●	○●
Unseen textual analysis	○●	○●	○●

Revision techniques

Although a large part of your study for Media Studies is production-based or takes the form of critical research essays, you also need to revise and prepare for exam papers. Media Studies, as you now know, is different from many subjects because it is based around your knowledge and application of key concepts rather than your learning of particular topic areas. In addition, only some of the units require essay-type answers, so you should check very carefully with your teacher to make sure you are revising the right things for your specification. In this section we give you some general advice about how to revise for Media Studies.

Basic revision techniques ●●●

For your essay questions, you need to start by making sure you know what your topic areas are. Collect together all your notes about the texts which you have been studying in relation to these topic areas and organise your revision time to make sure that you cover the following strategies:

→ Make sure that you have access to copies of any specific texts which you have studied in relation to your topic areas. If these are moving image texts, make sure you watch them again at least once. If they are print texts, make sure you study them at least once more.

→ Once you have watched or read your texts again, make sure you can deconstruct them in some detail. Make notes as you deconstruct them.

→ Identify any key features from this deconstruction which relate to the key concepts.

→ Identify any key features from this deconstruction which relate to your specific topic area for this unit (for example, specific aspects of framing or mise-en-scène for a documentary which help you easily define the documentary form being employed or use of specific generic conventions in a film).

→ Having revisited your primary texts, it might be a good idea to look at some other related texts. If you are revising a specific film genre try to find other films within the same genre which you can study. You could use a resource such as www.imdb.com to help you identify early films within your topic genre, for example.

→ Make sure that you keep detailed notes while you are watching/reading these texts so that you can compare them with your primary texts and use specific examples in your answers when using these secondary texts to support points you make in your essays.

→ Try to read around your texts. Read what reviewers and critics have to say about your texts, but remember that a lot of what they say is related to evaluating these texts in terms of whether they are 'good' or 'bad' rather than detailed deconstruction.

→ Make sure you have specific examples from your texts under each of the key concept headings in your notes so that you can use these in the essays where appropriate.

Reading around the topics ●●●

In both your unseen papers and your essay answers you will be rewarded for critical autonomy – in other words the ability to engage with a range of media texts in a meaningful way. The best way to develop this as part of your revision strategy is to extend your media consumption. In fact, it's probably a good idea to get into these habits as early as possible in your course so that you can build the habit and broaden your experience. The more experience you have as a media consumer, the easier autonomy will become but this does assume wide consumption, not simply reading or viewing what you like. This also means engaging with texts where you are not part of the target audience so that you can be aware of how different audiences are targeted and how different texts are constructed. So, for example:

→ Watch at least one news bulletin every day, listen to at least one radio bulletin and read one newspaper. Vary your choices and think about the application of the key concepts to these texts.
→ Read the TV listings carefully every day so you get a sense of what TV genres are currently popular and what channels are showing which programmes.
→ Visit the newsagent's at least once a fortnight to look at the range of niche magazines and identify the codes and conventions they use to target their specific audiences.
→ Try to watch or listen to critics such as film or televison critics.
→ Read relevant publications such as *Media Guardian* (published with the *Guardian* every Monday), *Broadcast* and *Sight and Sound* magazine. There are also specialist Media Studies magazines such as the *Media Mag* published by the English and Media Centre especially for A level students.

Media theories and debates ●●●

All the topic areas which you are likely to be writing about are based around some element of media theory. If you are studying documentary, for example, you will be expected to know about documentary form and the history of documentary film making. Almost all topic areas will relate to key theoretical areas such as audience theory, genre theory or narrative theory, so you need to make careful revision notes from your study notes in these areas.

There will also be key debates which will be relevant to your topic area. It may be that you will need to be aware of debates around violence in the media, say, or financing of British films.

Especially at A2 level, you must apply your knowledge of these key theories and debates in your answers. As with the key concepts, structure your revision notes into sections, e.g. background, basic theories, key terms or people and application to key texts. As always, make sure you can back up comments about these debates and theories with examples from your texts so you can apply these in your essays.

Critical essays

All the specifications ask you to complete critical research for one of your units. This gives you a chance to develop independent research skills which will be useful to you if you are going on to higher education. In this section we look at the process of critical research and how you can write about the process in an examination. This section links with the section on completing independent research.

Selecting your research area

It may be a requirement that you select your research topic from a specified list or that you have to choose a general concept to focus on, such as representation or genre. Alternatively, you may be able to choose your research area. Whatever the requirements, you should ensure that you choose a very specific area to study from within the topic. It is impossible to study a broad area such as 'representation of women' in totality and write in any detail, whether you are writing a critical essay or writing about your research in an exam context. A better study area would be 'representation of women in British Asian films' because it will allow you to research in far more depth.

Research methodology

As mentioned in the section about independent research, you will need to undertake primary and secondary research around your topic area and to record your findings carefully. Primary research (e.g. audience research), is not very helpful unless the findings are analysed. It's a good idea to use a combination of qualitative and quantitative research if you can, so that you can compare your findings.

Depending on your research topic, you should also identify a range of primary and secondary texts which you will use for your study. Make sure you check with your teacher that you have chosen appropriate texts to study.

Existing critical research material

Remember that you must demonstrate that you have undertaken research involving existing critical material. This will involve textbooks, Internet research, media theory and reading journals and magazines to get a range of critical responses.

Profile, expectations and reception of target audience

Another very important aspect of your research is, of course, your research about the target audience for your chosen texts. You should identify who the target audience are for your texts and remember that identifying the audience also means profiling them. It's not enough to say 'young people between 16 and 25' as that is very broad.

By exploring the texts in relation to audience expectations you will also be able to contextualise them in terms of genre and media context. Do the texts fulfil or challenge audience expectations? Remember, however, that this is not an evaluative exercise in terms of whether the audiences liked the texts, but in relation to genre and function.

Institutional factors

It's also important to evaluate the texts in their institutional context – for example, to consider how the institutional ideologies have been reflected in the texts and what key institutional factors have shaped the texts during construction. By considering the ideologies in this institutional context, it is much easier to define them and reflect on them in relation to the media issues which you are exploring.

Wider contexts

It is also essential when evaluating texts within a topic area to think carefully about the contexts of production and consumption as well as the institutional context. Ask yourself:

→ Was this part of a wave of similar texts?
→ Was it a groundbreaking text?
→ Did it provoke a lot of reaction or controversy?
→ Is it typical of its time and place?

Drawing conclusions

Once you have completed your research into your topic area you should be able to start to draw your conclusions. Questions you might need to think about as you do this are:

→ What is the relationship between your texts?
→ Why did you choose these texts for this topic area?
→ What texts have you not studied which would have been useful?
→ What are the key discoveries you have made about your topic area?
→ Have you got the results you expected from your research?
→ Which media theories have you applied during your research?
→ Why did you use these theories?
→ What media debates have you engaged with during your research? Why these debates?
→ Which was your most successful research method? Why?
→ Which was your least successful research method? Why?
→ How did you select your research methodologies? Why?

Critical Research Exam

If you are taking your exam with OCR, you will need to answer questions about your research methodologies and how you completed your topic research, so it is very important that you have thought about these questions.

Action point

Always investigate what other texts were produced in a similar genre around the same time as a text you are studying, so that you can draw some conclusions about trends at the time.

Watch out!

Remember that if you are asked about your research methodologies you must be able to write about the methods you have used.

Exam questions

Don't forget

Always read all the instructions on the exam paper – even if you think you know what they say. Things do change and these instructions are there to help you.

Examiner's secrets

It's very helpful if you list the primary texts you are using at the beginning of your essay as the examiner then knows what you are focusing on. Do give dates of publication or broadcast as well.

Watch out!

Remember that you won't get credit for using theories or debates unless you can apply them to your texts in the essay!

As with all subjects at AS and A2 level, you will have to write essay-type answers under exam conditions. Although the exact requirements will vary for each of the specifications, the basic process of writing a Media Studies essay is the same.

The basics

As with all essays, your answers should have a beginning, a middle and an end and you should make sure that you support your points with examples from your texts. You should always make sure you plan your answer before you start. Make sure you read the question properly and don't fall into the trap of reading it too quickly and writing an answer which is not focused because you have written the answer to a similar question which you wrote in your mock exam!

Media Studies essays

There are some big differences between essays for Media Studies and for other subjects and it is important to remember these.

Unlike subjects like English, there are no set texts for Media Studies. You will be studying topic areas such as Sitcom or Documentary or Crime and Media. Within that topic area, the choices of texts to study are left open. Your teachers will have chosen texts for you to study and you should know those texts very well by the time you sit the exam. However, the examiner will not know what texts you have studied, so you need to make this clear. Use lots of details from your texts, related to relevant key concepts, and don't forget to identify your texts. It can be very helpful to write the names of your primary texts at the top of your answer.

Essay titles are often very general for Media Studies, e.g. 'In what ways does a film's genre satisfy audience expectations?' It's very easy to answer these questions in very general terms with a few loose references to the texts that you have studied. However, strong answers will answer the question while keeping a very strong focus on detailed textual exploration. So, a good answer to this title would perhaps start by identifying a key film genre and two particular films within that genre which exemplified the generic codes and conventions. The answer would examine the relationship between audience expectations and these generic codes and conventions, exploring the debate about whether the institutions produce generic films, knowing that in this way they encourage the audiences to want more generic films and thus maintain demand, or whether the institutions are responding to audience demands for films within particular genres. This might lead on to debates about the Hollywood studios or bring in audience theory with discussion of effects theory and uses and gratifications. Strong answers would probably bring in additional texts to add more evidence, not just mentioning the films but using particular specific examples to support their points. There are, of course, other ways of approaching a discursive question like this – for example, through a discussion of genre theory or a detailed analysis of how global media organisations

such as Warner Bros control production, distribution and exhibition to such an extent that there is no opportunity for less formulaic blockbusters, in comparison with attempts by independent film makers to create innovative and non-generic films.

Critical autonomy ●●●

This is a very important concept in Media Studies. Basically it means **your** ideas and opinions, not someone else's. Good answers show very clearly that they are autonomous. This may be as basic as stating 'I think . . .', but usually critical autonomy is shown by the quality of your debate and the ability to really engage with the question. For instance, an answer about lifestyle magazines for teenagers would debate key issues relating to the representations and lifestyle aspirations used in such magazines in relation to a possibly vulnerable audience and write about the way such magazines pressure teenagers into conformity with particular constructions of an ideal lifestyle.

You can develop critical autonomy in lots of ways:

➔ Join in discussions in class so that you can develop your ability to sustain an argument.
➔ Read relevant articles in magazines and newspapers, so you know what current debates are and so you know what is happening in various media organisations.
➔ Use archives such as the *Media Guardian* webarchive, which has all sorts of articles such as articles about PSB broadcasting, newspaper institutions or developments in reality television, for example.
➔ Watch television programmes and listen to radio programmes which focus on media-related news – not showbiz gossip, but programmes which raise important issues.
➔ Watch specialist programmes which will help you to develop a critical sense for example *Film 200?*, or *Liquid News*.

General dos and don'ts ●●●

➔ Always take time to plan your answer – think carefully about what the question is about, relate your answer to that question – and make sure you finish your essay!
➔ Identify the primary texts that you are using and make sure you back up your points with evidence from your texts.
➔ Think carefully about which of the key concepts and which of the media theories/debates you have studied you can use in your answer.
➔ Remember that there are no 'right answers' in Media Studies. Credit is given for critical autonomy, not for learned answers.

Action point

Why not organise a rota in your class to review a range of sources such as *Media Guardian* and report back to the rest of the group each week about what important issues have been raised?

Check the net

There are several web services which you can subscribe to which will send you emails daily or weekly about issues and debates. For example, *Media Guardian* will send a daily email about media-related stories in all the newspapers.

Independent research

All the specifications anticipate that you will be involved in some form of independent research during your course. This is at A2 level because it is assumed that your research will move beyond simply analysing texts and applying the key concepts to these texts. You will need to research a particular topic such as Crime and the Media. You may be able to choose your research area or you may have to select from a list. Either way, you will need to identify key texts to explore and decide how you will research your topic area.

Primary and secondary rsearch ●●●

Primary research is first-hand research. So, primary research in relation to a particular film genre would start with watching a range of films within the genre, perhaps from different periods in the history of the genre and different countries of production. Primary research into violence in computer games would require playing a range of games which have been described as violent – and probably some which had not, for comparison. Obviously, it is a good idea to obtain copies of your primary texts, as a single viewing, playing or reading is unlikely to be enough. It helpful to cover a range of texts during your primary research phase. Although you may only use some of your texts as primary texts you will also need some secondary texts as additional, supporting evidence.

Be aware that some specifications ask for primary texts to be 'contemporary' and this can be defined very exactly. If you are at all unsure about your choices of texts, you should ask your teacher.

Whatever your chosen concept area, it's a good idea to try and find a range of texts to study, and to ensure that you have got good coverage of the wider contexts of production – i.e. historial, cultural, social, cultural, economic and political. So, if you were looking at women and film, for example, you would initially start with films by women directors, texts which historically have affected representations of women on film, or roles available to actresses. You would want to compare national contexts as well as historical contexts and probably different economic and/or cultural contexts as well. So, once you had completed your primary research phase, you would be able to focus on a central issue, e.g. funding of films made by women directors across these differing contexts, supported by detailed study of other related areas which would probably influence funding as well (such as female representations within these films).

Secondary research really means second-hand research. In other words, this is research such as reading critical theory books related to your concept area or doing Internet research to find out more about your chosen research topic.

Audience research is also a form of secondary research, unless it is the primary focus of your research topic. If you are completing an

audience research analysis, remember that the results of your research are usually more interesting than the process of getting the information.

Qualititative and quantitative research ●●●

There are two basic forms of audience research which you can undertake.

Qualitiative research is just that. You research audience reactions to texts in terms of the quality of their response. Qualitiative research usually involves completing interviews (group interviews and personal interviews), writing down everything which is said during the interviews and collating this information to draw conclusions. This can be a good way to explore audience responses to texts – for example, if you were researching children and televison and wanted to find out how a particular age group or gender responded to a particualar text. Qualitative research allows you to examine how a particular audience receive a particular text.

Quantitiative research draws conclusions based on numbers of responses. A questionnaire is a typical form of quantitatitve research. By asking 100 people to complete a questionnaire about their attitudes to and experiences of the 'cult of the celebrity', say, you may be able to draw some conclusions about general attitudes.

The big problem with questionnaires can lie in the construction. A poorly constructed questionnaire may end up providing little useful information. More sophisticated forms of quantitative research include the types of research carried out by BARB, for example, collecting information about how many people watched a particualar television programme. This type of research will not identify whether this was active or passive consumption but simply whether they were present while that programme was on.

Secondary research sources ●●●

There are various places where you can get secondary source material relating to your research area and chosen texts. There is some more information in the next section about using the Internet for this research but some good sources are listed here.

→ Your teacher may have access to books and articles which they can lend you as well as other texts you can use.

→ Your school or college library should have some relevant theoretical books which may have sections relating to your topic area.

→ Your local library will have a range of texts which you can use.

→ As a student, you should be able to get a day pass for your local university library which will have a lot of relevant textbooks and journal articles which you can use. You will be able to start your search by using the library catalogue (computerised) in the first instance and in most university libraries there is always someone available to help you.

→ There is a range of material available on the Internet and the next section suggests ways that you can find this information.

Links

You can find out more about the 'cult of the celebrity' on p. 134.

Don't forget

The most important form of primary research you can do is reading or watching the text! The best place to start with secondary research is to ask your teacher for advice.

Using the Internet for research

Probably one of the most important tools you have to help you when you are researching for Media Studies is the Internet. One of the problems you often have is that many of the texts and debates that you are researching are contemporary, which means there are not that many sources you can use. Many of the best contemporary sources are on the Internet but, as you know, the problem with the Internet is finding the information.

Using search engines effectively

Probably the search engine which you will use most is Google. There are other very good search engines but this is the one that people seem to use most. Other useful search engines include Kartoo and Dogpile. However, the thing that will make the most difference for you is **how** you search.

Example – You are looking for research material about the relationship between celebrities and the tabloid press.

1. The simplest way to do a search is to use the key search words for what you are looking for: *celebrities, tabloids.*

   However, this brings up over 22 000 documents for you to sift through.

2. If you want to focus your search down a little more, you can use speech marks to force Google to search for a phrase as a whole, not just the individual words. So you would use '*celebrities and the tabloid press*' as a phrase.

3. If you still can't find useful information you can try searching for particular types of information. For example, if you are looking for academic articles to read about this topic, you could use the search string ('string' just means 'list') *celebrity, tabloid, .ac* and the only sites that Google would search would be schools and colleges (as they are the sites with the .ac in them).

4. If you are researching articles relating to a particular celebrity story in the tabloids (remembering that you need to use detailed examples in your research!) you might use + and – in the search string. So, to make sure a word had to be included in the search, you would use + and to make sure Google avoided pages with a particular word you use –.

   For example, if you were intending to investigate the relationship between Victoria Beckham and the tabloid press, you might well want to look at a range of material which did not involve David Beckham as well. So your search string would be *celebrity, tabloid, +Victoria, – David, Beckham.*

   By using a very specific search string in this way, you will find a lot of very useful articles about Victoria Beckham's relationship with the tabloid press but without having to sift through lots of articles about David Beckham as well. If you wanted to focus on a particular tabloid, it would be easy to add *+Sun* to the string, for example, to find a range of material very quickly.

Check the net

Some of the other search engines you can try are:
www.dogpile.com
www.kartoo.com
www.yahoo.com

Watch out!

To make sure you don't get too lost when searching for information, if you hold down 'shift' each time you click a new link, it will open in a new window so you can keep a lot of information on the screen at once.

Using Internet research materials ●●●

It is very easy to print off a range of information relating to your study area, from a good range of sites. However, if you are using this information in your research you need to annotate the pages to show that you have used the information. You also need to make sure that you acknowledge Internet sources in your bibliography. The preferred way to reference websites is to give the full URL of the site and the date you last accessed the site. This is because, if you were looking at this piece of work at a point in the future, a date of last access for the site might help date the evidence or give an indication of whether the site would still be operating.

Information gateways

There are a lot of information gateways available which can help you locate particular information. Academic gateways, for example, database information about all research articles printed within a particular subject area so that you can search the database to locate the information you want.

For access to some of these gateways you will need to be using a college or school computer as access is only allowed from a machine on an academic network. If you can't get access to a particular gateway which looks helpful, ask your teacher or librarian if they can help you. A good starting point can be www.sosig.ac.uk.

Evaluating resources

There is a vast amount of information available on the web – far more than in your textbooks. However, not all of this information is accurate or useful so you have to evaluate it before using it for your research. Some questions you should think about are:

→ Who wrote the information on this site?
→ How old is the information?
→ Is this an independent site or does it have a particular institutional context?
→ Is it possible to cross check this information against another site?
→ What representations and ideologies are being constructed?
→ Does this affect the way the information is received?

Plagiarism

Perhaps the biggest danger with making use of a lot of Internet research is that it can tempt you into plagiarism. Plagiarism is very easy when you can simply copy and paste from a range of websites. However, plagiarism still counts as cheating and if you pass off someone else's work as your own you will not be allowed to enter it. Of course, you can use quotes providing you reference them, but you must not imply that someone else's words are yours.

Unseen textual analysis

All the exam boards require you to do some form of unseen textual analysis. This may be print materials, moving image materials or other materials. You may be shown texts from a genre you have studied. You may have to write about one text or more than one text. You will need to check with your teachers to make sure you know exactly what you have to do for your exams.

So why do we ask you to do unseen textual analysis? The answer is very simple – it's the best way you can demonstrate to us that you understand how a media text is constructed and why it is constructed that way. It's hard to prepare for an unseen, so you have to think on your feet.

The basics

First and foremost, you need to centre your answer around the key concepts. What you have to avoid is being descriptive. It's very easy to drift into writing a description of the text or texts unless you keep your focus.

Earlier in this book we identified how you can use MIGRAIN to help you structure your thoughts about the texts or texts. What you then need to do is try and organise your notes into a logical order. Below are some of the basic questions you should ask yourself while you are ordering these notes, ready to write your answer.

What is the medium, genre and form of the text(s)?

You should be able to identify these pretty quickly from the material you are given and this will enable you to start to frame your answer.

→ The **medium** will obviously shape the text in that it controls the way meaning can be created. A radio text is shaped differently from a print text.

→ The **genre** of the text will be communicated by the use of generic codes and conventions and by identifying the genre you can start to 'group' some of the media language used as generic conventions. This means you can write more succinctly and with good use of details about these as a group, rather than laboriously going through each section of the media language in turn.

→ The **form** of the text should also be fairly straightforward to identify. By identifying the form of the text, you again establish a group of conventions which you might expect and can address these as a group with examples.

What is the target audience for the text?

The next stage is to identify the **target audience**, because you can now start to relate much of what you have already noted about generic codes and conventions, for example, to this particular audience. Again, by thinking about how you can relate your analysis of the **media language** to another key concept, you can make more detailed points

Links

Look back at page 20 to remember how to use MIGRAIN to structure your answers.

Links

Remember that if you are not sure what genre is, you can look back at page 18.

Action point

For some texts, the target audience can be very broad and so this may not help you structure your answer in very much detail. Other texts can have very specific target audiences.

and also make them efficiently. Just remember to evidence each and every point you make with an example!

Ideology, representation and institution

Having established how the text conveys meaning and who the meaning is for, it is important to consider **why** the text is constructed that way. So you will need to look at the **institutional context** of production, the **ideologies** being established and the **representations** employed to convey these ideologies.

→ The first of these should be the **representations** of place and people since these will convey a great deal of information about the ideologies of the text and the expectations of the target audience.

→ Having identified target audience, genre, medium and form, you next need to examine the **institutional codes and conventions** being employed by the text. By keeping your focus on identifying institutional codes and conventions, you don't fall into the trap of identifying the institution for the text without examining how this shapes and influences the text!

→ From an identification of institutional context in relation to the target audience, you should be able to make some fairly specific comments about the **values and ideologies** being established by the text – how these reflect the institutional context, target audience and audience expectations.

At A2 level you will need to broaden this analysis to explore the wider contexts of production and consumption and also to consider media issues and debates raised by the texts used.

Wider contexts

You need to write in detail about the ways in which

→ economic contexts (e.g. available finance to make the text)
→ historical contexts (e.g. the technologies available to make the text)
→ political contexts (e.g. influences upon ideologies constructed)
→ cultural contexts (e.g. fashionable television genres of the time)

influence and shape the text.

Media issues and debates

It is unlikely at this level that you will be given texts to discuss which do not relate to some of the significant media issues and debates you have covered during your course. For example, your text may be violent and perhaps have been censored for some years, in which case you can write about effects theory and also about desensitisation in relation to cultivation theory. If you have prepared for your unseen and take a moment to think you will almost certainly be able to come up with several key issues and debates raised by the texts and to write about these – but don't forget to give evidence from the text to support what you say!

Watch out!

A lot of candidates don't write very much about explicit or implicit ideologies in a text and yet this is a very important area of analysis.

Examiner's secrets

At A2 level it is very important that you contextualise what you have to say. We assume that you can deconstruct the texts at this level!

Watch out!

It's a good idea to include some media theories and debates in your answer but don't be tempted always to use Propp and do remember that you don't need to explain the theories as well.

Glossary

Art house cinema – form of independent film which consists of low-budget films which are produced outside the mainstream

Auteur – a director of a film who is regarded as directing with a distinctive and creative style

BARB – the Broadcasters' Audience Research Board – an independent organisation which measures ratings for television programmes

BBFC – British Board of Film Classification

Bias – a preference that interferes with objective judgements

Binary oppositions – narratives are often shaped in terms of 'binary oppositions' – good *v* bad or old *v* young, for example

Bollywood – a nickname for the Indian film industry

Broadsheet – a physically large newspaper with a focus on hard news and political commentary

Camera angle – the height of the camera relative to the subject. In a high-angle shot, the camera is above the subject; in an eye-level or flat shot, the camera is at the same height as the subject; in a low-angle shot, the camera is below the subject

Code – a system of signs – verbal, aural or visual – used to convey meaning

Continuity editing – joins a number of shots taken at different times and even at different places so that they construct a narrative

Convention – a generally accepted custom or an established rule

Convergence – the coming together of different technologies or organisations

Critical autonomy – the ability to engage with a media text and have your own opinion about the text

Cultural imperialism – the domination of one culture by another through exposure and status

Cut – the change from one shot to the next

Cyberspace – the virtual environment created by the links among computers

Deconstruct – to take apart, or break down a media text in order to understand how it conveys meaning

Diegetic/non-diegetic – sound is diegetic if it is 'in the frame' – in other words; if it is happening in the narrative. It is non-diegetic if it happens 'outside the frame', e.g. a voiceover in a documentary

Discourse time – the time that the action of a narrative takes, which may not be the same as real time

Docudrama – a drama-based documentary that mixes documentary footage, focuses on real events and uses live interviews, etc, with dramatic scenes or re-enactments

Documentary – a factual film about a particular subject or a radio programme which treats an event or issue in depth

Dominant culture – the main culture of a society, i.e. the group of people with power or authority

Editing – the process by which film is cut into discrete shots and then joined with other shots

Editing (print media) – a process of selecting and altering writing in newspapers, magazines, and other print media

Ellipsis – ellipsis is an editing technique which joins two shots, missing out a passage of time between the two, e.g. leaving home and arriving at a destination, without including the journey

Encoding/decoding – Messages are created, or encoded, by media producers with a preferred meaning in mind, and then consumed, or decoded, by a receiver or audience. There are always multiple ways of decoding media messages: dominant, negotiated and oppositional

Enigma – one of Barthes' codes describing the way a text establishes a mystery for an audience which will be resolved by the narrative

Episode – one programme in a television series

Establishing shot (ES) – generally a long shot or extreme long shot used to show the setting

Extreme close up (ECU) – a close-up in which a small detail fills the entire screen

Extreme long shot (ELS) – a very wide shot, often a panoramic view

Form – the way a text is put together to create meaning for an audience

FPS – first person shooter – a typical computer game genre, e.g. *Halo, Combat Revisited*

Gatekeepers – the people who determine what will be printed, broadcast, produced, or consumed in the mass media

Genre – a way to classify a media text as a type

Hegemony – a theory of ideology as ideas, beliefs and values established by the dominant culture in society

High-angle shot – a shot in which the camera looks down upon the subject from a higher vantage point

Hyperlink – a word, phrase or image on one webpage that connects to another webpage

Icon – a highly valued person or product in popular culture or a visual representation of a particular object/concept

Ideology – a system of beliefs that is characteristic of a particular group

Interactive – an interactive media text requires an active consumer who must do things to engage with the text, not be passive

Internet – the interconnection of computers around the world so that they are capable of communicating with one another

Intertextuality – a postmodern term to describe the way we understand a postmodern text by its relation to other texts

Line of action – the line of action (also referred to as the '180-degree line') is an imaginary line drawn through the scene (for example, two characters conversing). To maintain continuity, the camera stays on one side of this line even as it focuses on different parts of the scene

Logo – a distinctive identifying symbol for communication, publication, or screen presentation

Long shot (LS) – a wide shot or a scene, which relates the subject to the background or setting

Low-angle shot – a shot in which the camera looks up at the subject from a lower vantage point

Mainstream media – media created for large numbers of people in society (related to popular culture)

Mass media – media such as television, radio and newspapers that can reach a large group of people

Media effects – a general term that refers to the impact of media consumption on individuals, society and culture

Media event – a reported event that appears to be news, but is actually staged by an individual, group or government

Media literacy – an understanding of media methods and messages

Medium shot (MS) – a medium close shot in which the head, shoulders and chest are normally included in the frame

Mise-en-scène – this term refers to the way a shot is visually staged for the camera. It literally translates as 'what's put into the scene', and to describe a shot or a scene's mise-en-scène is to describe all the elements that go into composing its 'look'

Montage – 'the collision of shots'; bringing two images together to create a meaning that is not immediately apparent in either one of them alone

Moral panic – a moral panic is often started by the media when a person or events seem to be threatening the dominant hegemony

Motif – any repeated element in a film, but usually taken to mean some element that takes on special meaning through its repetition. Good examples are musical themes that are associated with certain characters or certain moods and which pop up whenever that character appears or that mood is supposed to be invoked

Narrative – the structure of a media text; the 'story' of the text. Narrative theory is a way of describing that structure

Network – a system of computers connected together; a company that distributes programmes for broadcast

News values – the criteria by which news stories are selected and constructed for broadcast or print

NRS – the National Readership Survey is the main organisation which collects information about circulation figures for newspapers and magazines

Pan – short for 'panoramic' shot; a horizontal movement in which the camera is turned from the left to right or right to left to follow a moving subject or give a sweeping view across a scene

Point of view (POV) – a subjective shot that shows the point of view of a particular character

Popular culture – widespread, prevalent, and current trends or fads; arts, customs, beliefs and all other products of human thought made by and favoured by large numbers of people in a society

Prime time – the most popular time to watch television, usually between 7 p.m. and 9 p.m.

Propaganda – media texts that are intended to convince people to accept or reject a certain idea

PSB – Public Service Broadcasting – broadcasting which 'entertains, educates and informs' all groups in society and is not made for profit

Ratings – the measurement of a radio or television station's audience size

RAJAR – Radio Joint Audience Research is responsible for collating radio audience figures

Shot – reverse shot – a series of shots which keep one character on the right side of the framing, looking left, and the other character on the left side of the frame, looking right. The sequence usually uses over-the-shoulder shots (in which one character is shot from behind, so that the back of their headand their shoulder appears on the side of the shot) to link

Sound editing – stage in which sound effects and music are added to the programme

Sound effects (SFX, FX) – sound from any source other than voiceover narration, lip-sync, or music

Split screen – an effect in which two or more shots occupy different portions of the scene at the same time

Stereotype – a simplistic and often two-dimensional representation of a group or place

Synergy – the establishment of a relationship between different media organisations which benefits both organisations

Tabloid – technically, a publication usually half the size of a standard newspaper page; an image-led newspaper with a focus on soft news

Target audience – specific groups of people that media producers or advertisers want to reach

Virtual reality – a computer-created world that seems very real

Voiceover (VO) – a voice or commentary recorded for use on a film's soundtrack

Zoom – shortening or lengthening the focal length of the lens to give a closer or wider view. This also affects the figure–ground relation, making the background appear closer or further away

Index